Present Situation and Development of Chinese Environmental Protection

CHINA ENVIRONMENTAL PERFORMANCE EVALUATION REPORT

中国环境保护现状和发展

——OECD 中国环境绩效评估中期报告

赵学涛 彭 菲 杨威杉 马国霞 编著

中国环境出版社·北京

图书在版编目（CIP）数据

中国环境保护现状和发展：OECD 中国环境绩效评估中期报告 / 赵学涛 等 编著 . —北京：
中国环境出版社，2013.12
ISBN 978-7-5111-1307-8

Ⅰ.①中… Ⅱ.①中… Ⅲ.①环境管理—评估—研究报告—中国 Ⅳ.① X32

中国版本图书馆 CIP 数据核字（2013）第 027489 号

出 版 人	王新程
责任编辑	俞光旭　王素娟
责任校对	唐丽虹
装帧设计	金　喆

出版发行　中国环境出版社
　　　　　（100062　北京市东城区广渠门内大街16号）
　　　　　网　　址　http://www.cesp.com.cn
　　　　　电子邮箱　bjgl@cesp.com.cn
　　　　　联系电话　010-67112765（编辑管理部）
　　　　　发行热线　010-67125803，010-67113405（传真）

印　　刷	北京中科印刷有限公司
经　　销	各地新华书店
版　　次	2013年12月第1版
印　　次	2013年12月第1次印刷
开　　本	787×960　1/16
印　　张	12.5
字　　数	290千字
定　　价	68.00元

【版权所有。未经许可，请勿翻印、转载，违者必究】
如有缺页、破损、倒装等印装质量问题，请寄回本社更换

前言

改革开放以来，中国经济社会发展取得了举世瞩目的成就。1978—2010 年，中国 GDP 年均增长 9.8%，经济总量由世界第十上升为世界第二，综合国力明显增强，人民生活水平显著提高。但是，中国在经济增长方式方面，存在着高投入、高消耗等问题。因此，转变经济发展方式，实现绿色发展，这不仅是中华民族长远发展的战略性选择和必然需要，也是对全球可持续发展的积极贡献，将对人类发展产生重要影响。

经济合作与发展组织（OECD）作为绿色发展的倡导者之一，提出了绿色发展的概念性框架及监测评估工具，尤其是自 1992 年开始的环境绩效评估，对于帮助参评国家认识自身在经济发展过程中所面临的环境问题并提出对策，推动能源资源利用效率提高和环境管理政策的实施效果起到了积极作用。2005 年 10 月原国家环境保护总局和 OECD 签署了双方合作开展中国环境绩效评估的协议，并于 2006 年 11 月在北京发布《OECD 中国环境绩效评估报告》的结论和建议部分，2007 年 7 月 17 日在北京发布了该报告的全文。报告回顾了中国 10 年来在环境保护方面作出的努力，认为中国目前环境保护存在的突出问题是环境政策实施的有效性和效率亟待提高，并建议提高各级环保部门的监测、监督和执法能力，扩大各种经济手段的使用范围以及强化地方政府的环境责任，进一步将环境因素纳入经济综合决策。最后，针对环境管理涉及的各个领域，提出了 51 条意见和建议，以帮助中国在可持续发展中加强环境绩效。

中国政府已经深刻认识到实施绩效管理对于提高政府运作效率，加强政策实施效果有积极意义。尤其是在环境领域，自"十一五"以来，中国政府将化学需氧量（COD）和二氧化硫两项指标作为约束性指标，首次纳入国民经济和社会发展规划纲要，表明国家在环境保护的认识上发生了重要变化。从实施效果来看，截至 2010 年底，"十一五"环境保护目标和重点任务已全面完成，COD 和二氧化硫排放量分别比 2005 年下降 12.45% 和 14.29%，超额完成了规划目标任务。同时，污染治理设施快速发展，城市污水处理率由 2005 年的 52% 提高到 72%，火电脱硫装机比重由 12% 提高到 82.6%。与此同时，环境质量状况有所改善，全国地表水国控断面水质优于三类的比例提高到 51.9%，全国城市空气二氧化硫平均浓度下降 26.3%。这些成绩的取得正是"十一五"期间中国政府强化环境规划和政策实施、加强环境监管基础能力、综合运用多种政策手段的结果。环境保护作为转变经济发展方式、推动绿色发展和生态文明建设的重要抓手，在经济社会全面协调可持续发展中的作用显著增强。

为了进一步推动 51 条意见和建议的落实，环境保护部污染物排放总量控制司针对 OECD 提出的 51 条建议，以环境保护部环境规划院为主要技术支持部门，历时 3 年编写完成了《OECD 中国环境绩效评估 51 条意见和反馈》（以下简称《反馈意见》），先后两次征询了环境保护部各司局及发展改革委、住建部、国家林业局、商务部等十多个部委意见。为全面和充分反映中国在"十一五"期间所取得的环境绩效，在《反馈意见》基础上，环境保护部污染物排放总量控制司组织环境保护部环境规划院编写了《中国环境保护现状和发展——OECD 中国环境绩效评估中期报告》，报告在组织框架上借鉴了 OECD 提出的压力 - 状态 - 响应（PSR）框架，分别从环境压力和发展目标、中国政府作出的努力、发展成效、结论和展望，对中国政府在"十一五"期间的环境保护成效进行了全面回顾。

2012 年 10 月 10 日至 12 日，应经济合作与发展组织（OECD）邀请，以环境保护部环境规划院王金南副院长为代表的中国代表团一行 6 人赴法国巴黎参加了 OECD 环境绩效工作组会议，中国代表团就 2007 年以来 OECD 中国环境绩效评估工作完成后，特别是"十一五"中国环境保护工作取得的成效和颁布实施的重要环境政策，以及对 OECD 提出的 51 条政策建议的执行情况等与参会代表进行了交流和讨论。总体上看，这次会议是一个展示中国"十一五"环境保护成效的会议。OECD 对中国自 2007 年以来特别是"十一五"的环境保护取得的绩效给予了高度的赞赏和认可。

为进一步宣传中国政府在"十一五"以来取得的环境绩效，树立我国负责任的环境大国形象，在环境保护部支持下，我们结合《OECD 中国环境绩效评估 51 条意见和反馈》、《中国环境保护现状和发展——OECD 中国环境绩效评估中期报告》等材料，汇编成本书。本书有关材料的编写过程中得到了发展改革委、住建部、国家林业局、商务部等十多个部委的支持，他们对 51 条反馈意见从各自角度提供了翔实的材料，环境保护部各司局，尤其是污染物排放总量控制司于飞副司长、环境统计处毛玉如处长逐条审阅了 51 条反馈意见并提出了修改建议，国际合作司途瑞和副司长、刘宁处长和辜丽副处长对参加 OECD 环境绩效评估中期工作组会议提供了大力支持，办公厅钱勇处长对相关材料从完备性、表述的严谨准确性等方面提出了修改意见。此外，还要特别感谢环境规划院洪亚雄院长、王金南副院长对环境绩效评估工作的支持，王金南副院长率团参加了 OECD 环境绩效评估中期工作组会议，提出了出版本书的动议并就相关材料组织提出了指导意见，在此一并致以最诚挚的谢意。

本书编写组
2013 年 7 月

目 录

一、环境压力和发展目标 /1

（一）经济社会发展带来的环境压力 /1

（二）中国可持续发展总体目标与思路 /1

    1．中国推进可持续发展战略的总体目标 /1

    2．中国推进可持续发展战略的总体思路 /2

二、中国政府作出的努力 /2

（一）环保法律法规 /2

（二）环境标准和规划 /3

（三）环境管理体制机制 /5

（四）环保投入和能力建设 /6

（五）环境政策和措施 /6

    1．绩效管理 /6

    2．污染物排放总量控制 /7

    3．环境影响评价 /7

    4．排污收费和排污权交易 /8

    5．绿色信贷和保险 /8

    6．生态补偿 /9

    7．环保科技 /9

  8. 环境宣传和教育 /9

  9. 国际合作和履约 /10

三、发展成效 /11

  （一）大力发展循环经济，结构调整取得积极进展 /11

  （二）主要污染物排放降低，重点行业减排效果显著 /12

  （三）加强重点流域区域污染防治，局部地区环境质量得到改善 /15

  （四）全方位推进生态保护，生态建设成效显著 /17

  （五）环境保护能力建设取得显著成绩，环境监管能力得到加强 /19

  （六）信息公开工作更加规范，公众参与程度不断增强 /21

四、结论和展望 /22

  附件 OECD 中国环境绩效评估 51 条意见和建议落实情况 ……… 124

Preface / 75

1. Environmental stress and development objectives/79

1.1 Environmental stress as a result of economic and social development /79

1.2 China's objectives of sustainable development/80

2. Efforts of the Chinese government /81

2.1 Laws and regulations for environmental protection/81

2.2 Environmental standards and planning/81

2.3 Environmental management system and mechanisms/85

2.4 Input and capacity building for environmental protection/85

2.5 Environmental policies and measures/86

3. Development achievements/93

3.1 Develop circular economy with positive progress made in structural adjustment/93

3.2 Emission of major pollutants significuntly reduced and remarkable outcomes have been achieved in key industries/95

3.3 Strengthened pollution control in key basins and regions with improved environmental quality in local regions /99

3.4 Comprehensivcly promote ecological protection with remarkable outcomes in ecological construction achieved /101

3.5 Remarkable achievements has made capacity building of environmental-protection and enhancing environmental-regulation capacity /104

3.6 Information discloosure work performed in a more normative way and extent of public participation continuously strengthened/107

4. Conclusions and prospects/108

ANNEX:

ACTIONS TAKEN SO FAR AFTER THE
2007 OECD'S ENVIRONMENTAL PERECRMANCE
REVIEW OF CHINA..111

一、环境压力和发展目标

(一) 经济社会发展带来的环境压力

经济高速增长带来较大的资源环境压力。自 2000 年以来，中国经济步入快速发展轨道，国内生产总值从 2001 年的 109 655 亿元增加到 2010 年的 401 202 亿元，年均增长 10.5%。2008 年，中国国内生产总值超过德国，居世界第三位。2010 年，中国国内生产总值按平均汇率折算达到 58 791 亿美元，超过日本，成为仅次于美国的世界第二大经济体。在全球主要经济体大多面临负增长或停滞困境时，中国经济依然保持了相当高的增速并率先回升，为世界经济复苏作出了贡献。

人口素质、结构和分布问题是影响经济社会协调和可持续发展的重要因素。未来一个时期，人口数量问题仍然是制约中国经济社会发展的关键性问题之一，从总量看，中国仍是世界人口最多的国家，且每年以 5‰ 的自然增长率在增加，2010 年中国人口达到 13.4 亿人，占世界总人口的 19%。人口平均预期寿命由改革开放初期的 68 岁提高到 73.5 岁，达到中等发达国家水平。中国人口的持续增加，对资源环境产生较大压力。同时，中国仍面临较大的减贫压力，按照 2011 年中国发布的新农村贫困标准，扶贫对象尚有 1.22 亿，且大多生活在自然条件恶劣的区域，消除贫困的任务极为艰巨。

随着人口总量的持续增长，工业化、城镇化的快速推进，能源消费总量的不断上升，污染物产生量将继续增加，经济增长的环境约束日趋强化。环境容量和生态系统承受压力持续加大，同时也对环境管理、执法和监管能力提出较大挑战。

(二) 中国可持续发展总体目标与思路

1. 中国推进可持续发展战略的总体目标

中国推进可持续发展战略的总体目标是：人口总量得到有效控制、人口素质明显提高，科技教育水平明显提升，人民生活持续改善，资源能源开发利用更趋合理，生态环境质量显著改善，可持续发展能力持续提升，经济社会与人口资源环境协调发展的局面基本形成。

2. 中国推进可持续发展战略的总体思路

中国推进可持续发展战略主要包括五大内容：一是把经济结构调整作为推进可持续发展战略的重大举措；二是把保障和改善民生作为推进可持续发展战略的主要目的；三是把加快消除贫困进程作为推进可持续发展战略的急迫任务；四是把建设资源节约型和环境友好型社会作为推进可持续发展战略的重要着力点；五是把全面提升可持续发展能力作为推进可持续发展战略的基础保障。

"十二五"期间中国政府在强调经济社会发展的同时，强化了资源环境目标，提出了12项控制性指标，包括：耕地保有量保持在18.18亿亩。单位工业增加值用水量降低30%，农业灌溉用水有效利用系数提高到0.53。非化石能源占一次能源消费比重达到11.4%。单位国内生产总值能源消耗降低16%，单位国内生产总值二氧化碳排放降低17%。主要污染物排放总量显著减少，化学需氧量、二氧化硫排放分别减少8%，氨氮、氮氧化物排放分别减少10%。森林覆盖率提高到21.66%，森林蓄积量增加6亿m^3。

二、中国政府作出的努力

（一）环保法律法规

目前，中国在能源和资源管理、环境保护方面的法律法规总计有50多项，主要包括《中华人民共和国宪法》、《中华人民共和国环境保护法》、《中华人民共和国电力法》、《中华人民共和国煤炭法》、《中华人民共和国节约能源法》、《中华人民共和国可再生能源法》、《中华人民共和国矿产资源法》、《中华人民共和国水污染防治法》、《中华人民共和国大气污染防治法》、《中华人民共和国清洁生产促进法》、《中华人民共和国循环经济促进法》、《中华人民共和国固体废物污染防治法》等。

"十一五"以来，针对工业污染和能源领域，中国政府对环保法规不断完善，修订并发布了《中华人民共和国可再生能源法修正案》，制定实施《中华人民共和国循环经济促进法》，相继出台《规划环境影响评价条例》、《废弃电器电子产品回收处理管理条例》等 7 项环境保护行政法规。

（二）环境标准和规划

现行环境标准体系由两级五类标准组成，分别为国家级标准和地方级标准，标准类别包括环境质量标准、污染物排放标准、环境监测规范（环境监测分析方法标准、环境标准样品、环境监测技术规范）、管理规范类标准和环境基础类标准（环境基础标准和标准制修订技术规范）。围绕污染减排和环境质量改善，环保标准的数量以每年 100 项的速度递增，截至"十一五"末期，中国累计发布环境保护标准 1 494 项，现行标准 1 367 项，其中包括国家环境质量标准 14 项，国家污染物排放标准 138 项，环境监测规范 705 项，管理规范类标准 437 项，环境基础类标准 18 项。

需要强调的是，为改善空气质量，我国对空气质量标准进行了第三次修订，于 2012 年 2 月 29 日发布了新的《环境空气质量标准》（GB 3095—2012）。GB 3095—2012 对环境空气功能区分类方案进行了调整，取消了三类区；调整了污染物项目及监测规范；引入了 $PM_{2.5}$ 和 O_3 指标限值；收严了 NO_2、PM_{10} 等污染物的浓度限值，NO_2 的二级标准年均浓度由 $0.05mg/m^3$ 提高到 $0.04mg/m^3$，PM_{10} 二级标准年均浓度由 $0.1mg/m^3$ 提高到 $0.07mg/m^3$（表 2-1）；调整了数据统计的有效性规定；增加了指定实施的其他污染物限值附录。

同时，中国政府对饮用水标准进行了修订，新修订的《中华人民共和国生活饮用水卫生标准》（GB 5749—2006）规定了生活饮用水水质卫生要求、生活饮用水水源水质卫生要求、集中式供水单位卫生要求、二次供水卫生要求、涉及生活饮用水卫生安全产品卫生要求、水质监测和水质检验方法，其中对水质监测的指标由 35 项增加至 106 项，增加了 71 项；修订了 8 项。该标准于 2007 年 7 月 1 日实施，对提高供水安全提供了基本保障。

表 2-1 空气中污染物浓度限值　　　　　　　　　　　　　　　　　　　　单位：mg/m³

序号	污染物项目	平均时间	浓度限值 一级	浓度限值 二级
1	二氧化硫（SO_2）	年平均	0.020	0.060
		24h 平均	0.050	0.150
		1h 平均	0.150	0.500
2	二氧化氮（NO_2）	年平均	0.040	0.040
		24h 平均	0.080	0.080
		1h 平均	0.200	0.200
3	一氧化碳（CO）	24h 平均	4.000	4.000
		1h 平均	10.000	10.00
4	臭氧（O_3）	日最大 8h 平均	0.100	0.160
		1h 平均	0.160	0.200
5	颗粒物（PM_{10}）	年平均	0.040	0.070
		24h 平均	0.050	0.150
6	颗粒物（$PM_{2.5}$）	年平均	0.015	0.035
		24h 平均	0.035	0.075

为落实有关能源、资源和环境的法律法规和标准，中国政府建立了分级负责的规划体系，其中国民经济和社会发展规划纲要是国家层面最重要规划，该纲要对涉及中国国民经济发展的各个领域，包括经济发展、人口发展、社会保障、产业政策、资源和能源利用等以五年为周期，进行了全面的规划和部署，在各个领域均提出了具体的目标及相应的控制性指标。

为了形成人口、经济、资源环境相协调的开发格局，中国政府 2010 年发布了《全国主体功能区规划》，提出将中国国土空间按开发方式分为优化开发区域、重点开发区域、限制开发区域和禁止开发区域；其中：

➢ 优化开发区域重点是要加快转变经济发展方式，调整优化经济结构，提升参与全球分工与竞争的层次；

➢ 重点开发区域主要是在优化结构、提高效益、降低消耗、保护环境的基础上推动经济可持续发展，推进新型工业化进程，增强产业集聚能力；

➢ 限制开发区域主要是指农产品主产区，该区域重点是保护耕地，发展现代农业，增强农业综合生产能力，增加农民收入；

- ➤ 禁止开发区域是指重点生态功能区，主要是指在国土空间开发中限制大规模高强度工业化城镇化开发，以保持并提高生态产品供给能力。
- ➤ 规划还要求调整完善财政、投资、产业、土地、农业、人口、环境等相关规划和政策法规，建立健全绩效考核评价体系。

在环境领域，经过30多年的发展，我国已经基本形成了一套具有中国特色的环保规划体系。从规划层级上看，主要包括国家层面和地方层面的规划体系。其中，国家层面的规划由四个层次构成，第一层次是国家五年环境保护规划，它是国家总体规划，确定了国家层面的环境保护目标与指标、主要任务与措施；第二层次是国家环境保护专项规划，主要解决环境保护重点领域的突出问题；第三层次是由环保部门参与的有关环保的国家专项规划，体现了环境保护与资源开发利用及经济社会发展规划的衔接；第四层次是环保部门自身的发展规划，主要是为强化环保部门职责和能力建设而制定。地方层面的规划包括区域环境保护规划、省级环境保护规划和地市级环境保护规划三个层次。

"十一五"以来，随着国家和社会公众对环境与经济问题认识的逐渐加深，环境保护从认识到实践都发生了重要变化，环境规划正逐步摆脱过去那种从属于经济社会发展规划的困境，在贯彻落实国家环境保护战略，推动和实施环境保护"三个历史性转变"，探索实践中国环境保护新道路，增强环境保护工作参与宏观调控、优化经济增长等方面的作用大大增强。

（三）环境管理体制机制

为加强全国的环境保护工作，2008年3月，环境保护部正式成立，先后增设污染物排放总量控制司、环境监测司、宣传教育司、核设施安全监管司、核电安全监管司、辐射源安全监管司和环境保护部卫星环境应用中心等机构。新成立的环境保护部加大了总量减排、环境监管、核与辐射监管等重大问题的统筹协调力度，以环境保护部的成立为契机，各省先后将省环保局升格为环保厅。

为强化国家环境监管能力，2005年以来，原国家环保总局先后成立华北、华东、华南、西北、西南、东北 6 个区域环境保护督查中心；1999 年以来，先后成立北方、上海、广东、四川、东北、西北 6 个区域核与辐射安全监督站。其中各环境保护督查中心为环境保护部的派出机构，其主要职能是监督地方对国家环境政策、规划、法规、标准执行情况；承办重大环境污染与生态破坏案件的查办工作；承办跨省区域、流域、海域重大环境纠纷的协调处理工作。部分省市还成立了独立的区域督查中心，例如：江苏分别设立苏南、苏中、苏北三个区域环境保护督查中心、陕西成立陕北环境保护督查中心。

通过机构调整，进一步理顺了环境保护管理体制，基本确立了"国家监察、地方监管、单位负责"的环境监管体系。

（四）环保投入和能力建设

"十一五"期间，从中央到地方都加大了环保的财政资金支持。中央财政对环保工作的支持达到历史最好水平，累计下达预算资金 100.34 亿元，是"十五"期间的 4.71 倍。全口径中央环保投资达 1 564 亿元，是"十五"中央环保投资的近 3 倍。带动全社会环保投入 2.1 万亿元，有力地推动了环境保护基础设施建设。

环境保护部通过加强国家重点监控企业自动监控、污染源监督性监测、环境监察执法、环境信息与统计四个方面的能力建设，正在努力建立一套科学、系统的主要污染物排放总量数据传输、核定、分析体系，一套污染源监督性监测和重点污染源自动在线监测相结合的环境监测体系，以及一套严格的、操作性强的污染物总量减排考核体系。

（五）环境政策和措施

1. 绩效管理

为贯彻落实各项环境保护目标，在环境保护领域，中国建立了一系列绩效管理手段，在污染减排、城市环境综合整治、重点流域水质目标考核、生态环境质量评价等领域实施了目标责任制和绩效考核制度，强化了政府环境绩效管理。2011 年，中国政府启动了污染减排绩效管理试点，进一步深化了绩效管理工具在污染减排领域的应用。

积极推进城市环境综合整治与定量考核，全国661个城市全部纳入城考范围，2 000多万个城市环境管理基础数据全部实现网络报送和系统校核，采取省级互审互查和现场抽查等多种手段，并按照国际惯例采取电话入户方式开展公众满意度调查。

创建国家环境保护模范城市。这是一项具有中国特色的环境保护激励性政策，目的是通过模范城市创建，充分调动地方政府积极性，努力解决群众关注的突出环境问题，加大城市环保投入，强化环境基础设施建设，加速优化产业结构，加快城市环境质量改善。2010年，国家环境保护模范城市全年空气优良天数比例和地表水环境功能区水质达标率分别高于全国平均值21.91个和11.36个百分点，医疗废物集中处置率、生活污水集中处置率、生活垃圾无害化处置率分别高于全国平均值14.91个、24.74个、24.27个百分点。国家统计局调查显示，环保模范城市公众对环境的满意程度远高于其他城市。

2. 污染物排放总量控制

污染物排放总量控制是目前中国最重要的环境政策之一，该政策包括一系列的措施和手段。在国家层面，国民经济和社会发展"十一五"规划纲要规定了主要污染物排放控制的量化要求，为配合实现该目标，国务院专门制订了《节能减排综合性工作方案》，对节能减排工作进行总体部署。同时，为加强落实，中国政府将减排约束目标进行了层层分解，将减排目标落实到各级政府和各相关行业；在减排目标落实过程中，环境保护部采取了减排定期调度和核查制度，对减排目标的实施情况进行监督，并结合减排目标落实进展，采用"区域限批"、"挂牌督办"等手段，建立减排约束机制。同时，采取包括脱硫电价补贴等激励性措施，提高企业减排积极性。

3. 环境影响评价

2002年10月，第九届全国人大常委会第三十次会议审议通过了《中华人民共和国环境影响评价法》，确立了建设项目和规划的环境影响评价制度。2009年8月，国务院颁布《规划环境影响评价条例》，标志着环境保护参与综合决策进入了新的阶段。该条例要求国务院有关部门、设区的市级以上地方人民政府及其有关部门，对其组织编制的土地利用的有关规划和区域、流域、海域的建设、开发利用规划（以下简称综合性规划），以及工业、农业、畜牧业、林业、能源、水利、交通、城市建设、旅游、自然资源开发的有关专项规划（以下简称专项规划），应当进行环境影响评价。"十一五"以来，规划环评数量逐年递增，环境保护部共计完成158项重点领域规划环评的审查工作，各地完成超过2 300项规划环评审查。

4. 排污收费和排污权交易

为建立有利于环境保护的激励和约束性机制，我国不断完善排污收费政策，进一步规范了排污收费制度，扩大了排污收费覆盖面积。通过收严征收标准，加大征收稽查力度，排污费的环境效益逐渐加强。江苏、北京和天津等12个省份提高了二氧化硫排污收费标准，规范城市污水收费，提高污水收费标准，2009年以来，在统计的36个大中城市中，已有10个城市较大幅度提高了污水处理收费标准。

2009年以来，中国已在江苏、浙江等18个省市开展了排污交易试点，浙江、北京、山西、重庆等省市已经成立了省级排污权交易中心。2011年浙江省印发了《浙江省排污权有偿使用和交易试点工作暂行办法实施细则》用于指导排污权交易工作。《国务院关于加强环境保护重点工作的意见》中明确提出"十二五"期间中国将建立国家排污权交易中心，发展排污权交易市场。

5. 绿色信贷和保险

在绿色信贷领域，环境保护部联合中国人民银行、中国银行业监督管理委员会发布了《关于落实环保政策法规防范信贷风险的意见》、《关于全面落实绿色信贷政策进一步完善信息共享工作的通知》等文件。20多个省、市出台了实施性文件。江苏、广东等地开展了企业环境行为信用评价。环保部门和金融机构建立了良好的信息沟通和共享机制。4万余条企业环境违法信息、7 000多条环境审批信息已经纳入人民银行征信系统。2012年2月，环境保护部积极配合中国银行业监督管理委员会发布了《绿色信贷指引》，对银行机构实施绿色信贷做出具体规范和引导。

国务院2011年9月发布的《太湖流域管理条例》首次以国务院行政法规的形式，确立了环境污染责任保险制度。环境保护部联合中国保险监督管理委员会发布了《关于环境污染责任保险工作的指导意见》。上海、重庆等14个省（自治区、直辖市）积极开展试点，出台了实施意见。广西、湖南、河北、江苏等地区已开始实施高环境风险企业环境污染强制责任保险试点。据不完全统计，截至2011年底，全国已有1 800

余家企业投保，保费总额达到 1.02 亿元，保险限额 174.30 亿元。

6. 生态补偿

2008 年修订的《中华人民共和国水污染防治法》首次以法律的形式对水环境生态保护补偿机制做出明确规定。环境保护部印发了《关于开展生态补偿试点工作的指导意见》，联合财政部、国土资源部下发了《关于逐步建立矿山环境治理和生态恢复责任机制的指导意见》。河北、河南、江苏、山东、辽宁、陕西、四川等地推进重点流域、重要生态功能区和矿产开发生态补偿试点。为了进一步规范生态补偿机制，推动生态补偿工作发展，国家已将生态补偿立法工作提上议事日程，《生态补偿条例》已被列为国务院 2011 年立法工作计划二类项目，由国家发展和改革委员会牵头，环境保护部和其他有关部门共同参与起草工作。

7. 环保科技

中国高度重视并不断提高科技对环境保护的支撑能力，《国家中长期科技发展规划纲要（2006—2020 年）》设立了十六个重大专项，其中四个与环境保护有关。"十一五"期间，环境保护部门组织实施了污染源普查、环境宏观战略研究和水体污染控制与治理科技重大专项研究为主要内容的三大基础性、战略性环保工程。通过开展污染源普查，摸清了全国污染源状况，建立了污染源信息数据库；通过国家环境宏观战略研究，提出了中国未来环境保护战略和措施；通过国家水体污染控制与治理科技重大专项研究实施，初步构建了中国水污染治理和水环境管理两大技术体系，为国家水环境综合整治和饮用水安全保障提供了有力支撑。同时，通过重大科研项目实施，带动了包括科研院所、重点实验室等科技平台建设，集聚和培养了优秀人才，壮大了环保科研队伍。

8. 环境宣传和教育

为提高公众环保意识，提高公众参与环境决策的能力。中国政府借助各种媒介组织开展了环保宣传活动。环境保护部借助每年的世界环境日，积极开展有影响、有创意的主题活动，以此提高公众环境意识，鼓励公众积极参与环境保护。从 2005 年起，环境保护部在每年联合国环境规划署发布世界环境日主题的同时，发布当年的世界环境日中国主题。

中国政府及时回应社会舆论对环保热点的关注。针对媒体和网民高度关注血铅超标、空气污染、铬渣倾倒、水电开发、减排政策等诸多话题，环境保护部门以发布新闻通稿、召开新闻发布会、通报会等形式，由新闻发言人、有关司局负责人或邀请专家出面通报和解释，受到舆论好评。同时，鼓励并支持环保NGO发展。

环境保护部一直十分重视环保民间组织对环境保护事业的推动作用，并积极支持和发挥NGO组织的作用，近年来中国涌现了一大批在国内外具有影响的环保NGO组织，例如自然之友、地球村、公众环境研究中心等。环境NGO组织在中国的环保事业中也发挥着越来越大的影响。例如"地球村"已经从简单的社区卫生维护和垃圾分类发展到现在对政府有一定影响力的社团，其主要活动包括：建立绿色社区、培育生态乡村等。

积极开展环保职业教育，组织开展绿色学校、绿色社区和绿色家庭创建活动。开展面向社会的环境责任培训。从2010年起，每年举办一期企业环境责任培训班。

9. 国际合作和履约

随着综合国力不断上升，国际地位及影响与日俱增，中国的国际环境合作经历了从封闭到开放，再到逐步融入国际社会的历程。许多国际环境问题的解决也越来越离不开中国的参与，中国参与国际环境问题已从最开始的"被动应对"到"积极参与"再到进一步开始"发挥建设性作用"。

中国充分借助国际环境保护合作平台，积极引进先进环保理念、管理机制，借鉴经验教训，不断完善中国环境与发展国际合作委员会高层政策咨询机制，激发中国环境与发展动力，助推中国形成特有的环境管理制度体系。国家出台的很多重大环境政策、制度和法规都借鉴了美国、日本、欧洲等发达国家的成功经验和教训，避免走发达国家已经走过的"先污染，后治理"、"以环境污染换取经济增长"的老路，为积极探索中国环境保护新道路作出了贡献。

在国际履约领域，目前，环境保护部牵头谈判、履行的环境国际公约5项、核安全国际公约2项、议定书4项。2002—2012年十年间，新批准公约和议定书包括《关于持久性有机污染物的斯德哥尔摩公约》和《生物安全议定书》等共计4项。十年来，环境保护部积极、深入、有效地参与国际环境公约履约，取得了丰硕成果，为全球可持续发展作出了积极贡献。

三、发展成效

（一）大力发展循环经济，结构调整取得积极进展

"十一五"以来，中国政府已认识到综合环境经济决策对于保护资源环境和优化经济增长具有重要作用，中国积极探索新型工业化道路，把调整产业结构、发展循环经济、推进战略性新兴产业发展、改造升级传统产业作为重要途径，努力转变经济发展方式。以信息化促进工业化，以工业化带动信息化，推动制造业核心竞争力的提升。不断加快发展现代服务业，积极倡导绿色消费，逐步提高发展质量。同时，国家发展和改革委员会、环境保护部、国土资源部、水利部、农业部等各大部委之间建立了行之有效的工作机制，共同协调经济发展和资源环境保护之间的矛盾。

"十一五"期间，大力推进节能减排工作。全国单位国内生产总值能耗降低19.1%；同时，充分发挥污染减排倒逼机制作用，累计关停小火电机组7 683万 kW，提前一年半完成关闭5 000万 kW 的任务；淘汰落后炼铁产能1.2亿 t、炼钢0.72亿 t、水泥3.7亿 t、焦炭9 300万 t、造纸720万 t、酒精180万 t、味精30万 t、玻璃3 800万重量箱（见图3.1）。

大力发展循环经济。2005年，中国发布了《国务院关于加快发展循环经济的若干意见》，出台了相关财政、税收、投融资等政策，有效引导和支持循环经济发展。2006年，将循环经济关键技术列入《国家中长期科学与技术发展规划纲要》。2008年，发布了《中华人民共和国循环经济促进法》，这是继德国、日本后世界上第三个专门的循环经济法律。2005年以来，组织开展国家循环经济试点示范，先后确定了两批共178家试点单位。28个省（市、区）开展了省级试点，共确定133个市（区、县）、256个园区、1 352家企业作为试点，总结凝练出60个中国特色的循环经济典型模式案例。2010年，资源循环利用产业产值超过1万亿元，从业人数超过2 000万人；钢、有色金属、纸浆等产品1/5～1/3的原料来自再生资源，水泥原料20%来自固体废弃物，工业固体废物综合利用率达到69%。

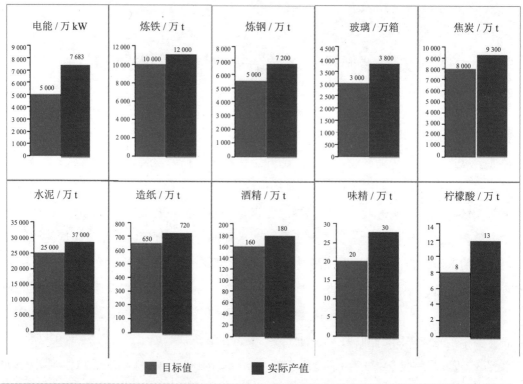

图 3.1 "十一五"期间落后产能淘汰情况

（二）主要污染物排放降低，重点行业减排效果显著

在"十一五"期间经济增速和能源消费总量均超过规划预期的情况下，二氧化硫减排目标提前一年实现，城市和工业领域化学需氧量减排目标提前半年实现。2010年全国城市和工业领域化学需氧量、二氧化硫排放总量分别较 2005 年下降 12.45% 和 14.29%，均超额完成 10% 的减排任务（见图 3.2 和图 3.3）。2010 年，全国造纸、化工和纺织行业化学需氧量排放强度比 2005 年分别下降 73.9%、66.7% 和 50%，电力、非金属矿物制品、黑色金属冶炼行业二氧化硫排放强度分别下降 72.5%、58.1% 和 50%。

重点行业二氧硫化去除率不断提高，电力行业二氧化硫减排效果显著。2010 年，工业二氧化硫去除率为66%，比2005年提高了30.5%，其中，电力生产、黑色冶金、化工、有色冶金、石化等行业的二氧化硫去除率分别为68.3%、31.9%、54.2%、89.9%、79%，比2005年分别增加50%、8.5%、5.5%、3.9%、23.3%（见图3.4）；"十一五"期间，占工业行业二氧化硫排放量半数以上的电力行业二氧化硫减排比例为22.9%，高于全国平均削减率。

废水排放重点行业化学需氧量去除率不断提高，造纸行业化学需氧量减排效果明显。2010年，工业化学需氧量去除率为80.2%，比2005年提高10.1%，其中，造纸、化工、农副食品加工业的化学需氧量去除率分别为82.4%、76%和76.7%，分别比2005年提高15.2%、21.7%和38.5%（见图3.5）；"十一五"期间，造纸行业化学需氧量减排比例为40.4%，高于全国平均削减率。

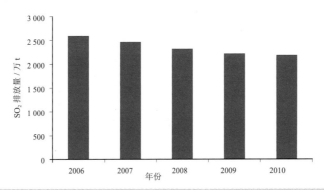

图3.2 "十一五"期间二氧化硫排放情况

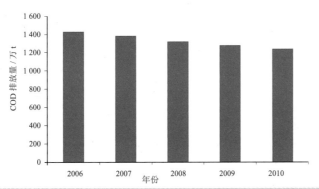

图3.3 "十一五"期间COD排放情况

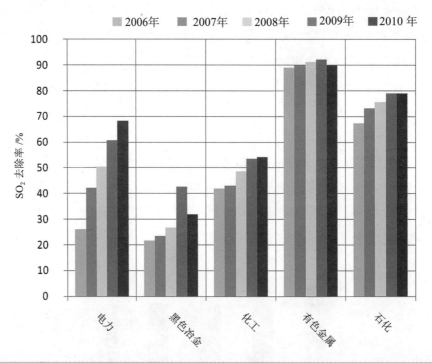

图 3.4 "十一五"期间重点行业二氧化硫去除率变化情况

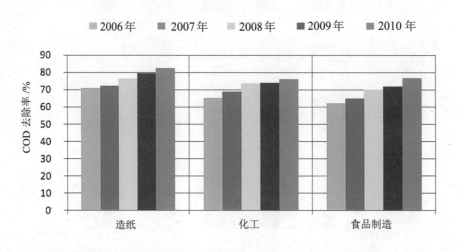

图 3.5 "十一五"期间重点行业化学需氧量去除率变化情况

（三）加强重点流域区域污染防治，局部地区环境质量得到改善

中国通过实施大气主要污染物排放总量控制制度，加强工业污染源治理，提高城市清洁能源消费比重和能源利用效率，强化机动车污染防治等各项措施，使得大气环境质量有所改善。"十一五"期间，中国城市空气质量达到二级标准的比例逐年提高，三级标准和劣于三级标准的城市比例呈下降趋势。达到二级标准城市比例由2006年的58.1%提高到2010年的79.2%，达到三级标准城市比例由2006年的28.5%下降到2010年的15.5%，劣于三级标准城市比例由2006年的9.1%下降到2010年的1.7%（见图3.6）。全国城市环境空气中二氧化硫、可吸入颗粒物的年均浓度分别下降26.3%和12%（见图3.7）。

2010年，全国地表水国控监测断面中，Ⅰ～Ⅲ类水质断面比例比2005年提高14.4个百分点，劣Ⅴ类水质断面比例下降6.6个百分点；国控重点湖库水体富营养化程度逐渐减轻，2010年26个国控重点湖泊（水库）中，满足Ⅱ类水质的1个，占3.8%；Ⅲ类的5个，占19.2%；Ⅳ类的4个，占15.4%；Ⅴ类的6个，占23.1%；劣Ⅴ类的10个，占38.5%。2005—2010年，113个环保重点城市地表水国控监测断面COD年均浓度由7.2mg/L下降到4.9 mg/L，下降31.9%（见图3.8）。

"十一五"期间，中国重度酸雨强度有所减少，发生较重酸雨（降水pH年均值小于5.0）和重酸雨（降水pH年均值小于4.5）的城市比例呈下降趋势，较重酸雨和重酸雨城市比重分别从2006年的10.7%和17.7%分别下降到2010年的8.5%和13.1%。

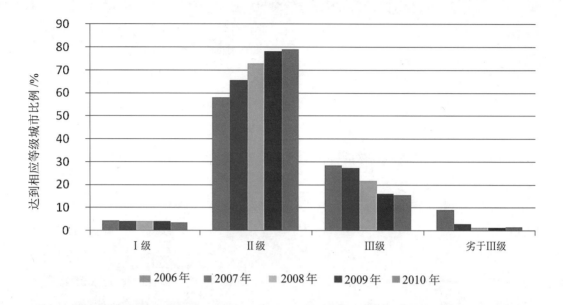

图 3.6　全国空气质量等级城市分布

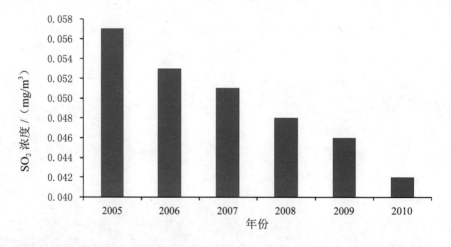

图 3.7　113 个环保重点城市二氧化硫浓度

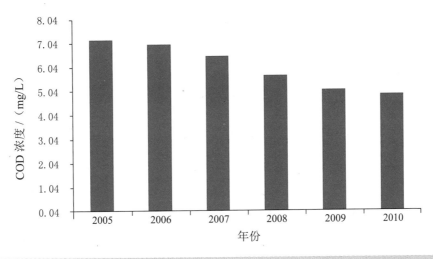

图3.8 759个地表水国控断面COD浓度变化

重点流域区域污染防治成效显著。按照《重点流域水污染防治专项规划实施情况考核暂行办法》的要求,全面建立了重点流域省界断面水质考核制度,重点流域水污染防治专项规划项目完成率为87%,比"十五"提高22.8个百分点,累计完成投资1 389亿元。继"绿色奥运"成功之后,圆满完成上海世博会、广州亚运会空气质量保障任务。

(四)全方位推进生态保护,生态建设成效显著

作为生物多样性公约的签约国,中国政府高度重视生物多样性保护工作。2007年发布实施《全国生物物种资源保护与利用规划纲要》(2006—2020年),2010年发布实施《中国生物多样性保护战略与行动计划》(2011—2030年)等规划和计划,提出了中国生物多样性保护的3个阶段目标、10个优先领域、30个优先行动和39个优先项目,成为中国生物多样性保护的纲领性文件。2010年,国务院成立了由李克强副总理担任主席的"中国生物多样性保护国家委员会",并组织召开生物物种资源保护部际联席会议。为掌握中国重要生物物种资源及生物多样性情况,2004—2009年,环境保护部

联合相关部门开展了全国重点生物物种资源调查,完成了相关物种编目和调查报告,掌握了中国重要野生生物资源、畜禽品种资源、药用生物资源、微生物菌种资源等的种群分布和保护状况,并建立了国家生物物种资源数据库和信息平台。2007—2012年,组织开展了全国生物多样性评价。至2011年底,全国各省区市全部完成了生物多样性评价工作,通过评价工作的开展初步掌握了全国各省区市的生物多样性现状、空间分布特征及主要威胁因素,整理形成了基于县级生物行政单元的生物多样性评价数据库,提出了各省生物多样性保护及可持续利用的对策建议。

自然保护区建设方面,2007—2011年,全国自然保护区由2 531个增加到2 640个(见图3.9),2011年自然保护区总面积约149万km^2(其中陆域面积约143万km^2,海域面积约6万km^2),陆地自然保护区面积约占国土面积的14.9%。

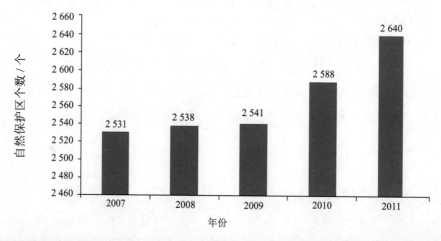

图3.9 自然保护区个数变化

林地、草原、湿地保护能力不断加强,沙化土地面积实现净减少。林地保护方面,近年来中国各级林业部门大力开展荒山造林、四旁植树,以及特色经济林、碳汇林、能源林基地建设,加强优质乡土和珍贵树种培育,2007—2010年,中国造林总面积由391万hm^2增加到591万hm^2,提高51.2%(见图3.10)。草原生态保护方面,截至2011年年底,各省区共实施草原禁牧面积8 066.7万hm^2;推行草畜平衡面积17 066.7万hm^2;

落实享受牧民生产资料补贴牧户198.7万户。湿地保护方面，2009年以来，中国启动了第二次全国湿地资源调查，初步建立了以湿地自然保护区、湿地公园为主的湿地保护网络体系。同时，中国认真履行《湿地公约》，强化中国履行《湿地公约》国家委员会的作用，顺利实施中澳、中德、中美等国际合作项目。截至2011年，中国新增湿地保护面积33万hm^2，恢复湿地2.3万hm^2，新增4处国际重要湿地和68处国家湿地公园试点。国际重要湿地达41处，面积为371万hm^2，湿地示范区面积达到349万hm^2。荒漠生态系统保护和治理方面，近年来，中国积极推进国家级沙化土地封禁保护区建设和区域性防沙治沙工作，完成了第四次荒漠化和沙化监测。"十一五"期间中国政府完成沙化土地治理面积1 081.41万hm^2，沙化土地由20世纪末的年均扩展3 436km^2变为目前的年均缩减1 717km^2，总体上实现了沙化土地面积净减少。

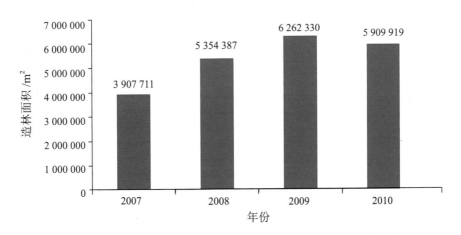

图3.10 造林总面积变化

（五）环境保护能力建设取得显著成绩，环境监管能力得到加强

环境基础设施建设突飞猛进。"十一五"时期，全国累计建成运行5.78亿kW燃煤电厂脱硫设施，全国火电脱硫机组比例从2005年的12%提高到82.6%；全国累计

建成城镇污水集中处理设施2 832座；日处理能力达到1.25亿 m^3；"十一五"期间增加约2 000座，新增污水处理能力超过5 000万 t/d，全国城市污水处理率由2005年的52%提高到75%以上。河南、江苏、浙江、广东等省县建成污水处理厂，宁夏在西北地区率先启动县县建设污水处理厂。（见图3.11和图3.12）

通过环境监管能力建设和环境保护基础工作的资金投入，建立健全了国家监察、地方监管、单位负责的环境监管体制，全国52%的县区级环境监测站基本达到监测设备标准化建设要求，环境监督执法力度不断提高。2006年以来，针对重金属污染、造纸企业、污水处理厂和垃圾填埋场等重点问题开展专项检查，全国共出动执法人员1 065余万人次，检查企业446万多家次，查处环境违法企业8万多家次，取缔关闭违法排污企业7 293家，停产治理企业5 981家，限期治理企业6 432家，挂牌督办环境违法案件1.9万余件，严厉打击了环境违法行为，维护了群众环境权益。

环境应急处置体制机制日益完善。环境保护部组建应急中心，建立环境应急专家库，全国有1/3以上的省级环保部门成立了专门环境应急管理机构。深化了部门应急联动机制建设，按照环境保护部与安全监管总局《关于建立健全环境保护和安全监管部门应急联动工作机制的通知》的要求，目前已有近20个省（自治区、直辖市）的环保部门与安监部门签署了应急联动机制协议，环保安监应急联动进一步深化。"十一五"期间，共处置突发环境事件912起，一批社会高度关注的重特大环境事件得到妥善处置。

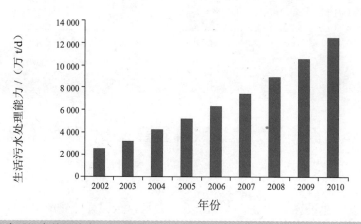

图3.11 城市生活污水处理能力

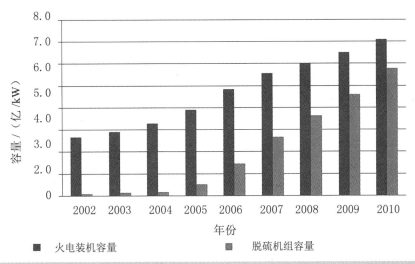

图 3.12　火电装机容量和脱硫机组容量

（六）信息公开工作更加规范，公众参与程度不断增强

2008 年 5 月 1 日，原国家环保总局颁布实施《政府信息公开条例》及《环境信息公开办法》，进一步规范和促进了政府和企业环境信息公开。中国政府加强了环保监管和环境信息披露工作。在信息发布方面，环境保护部逐年发布环境统计年报和环境质量公报，同时在环境保护部网站主页上适时公布污染源监测信息。各级环境保护部门也通过质量公报和媒体等手段，发布与环境质量相关的信息。

中国环保民间组织积极参与政策决策，共同推动环境保护事业发展。近年来，中国涌现了一大批在国内外具有影响的环保 NGO 组织，例如自然之友、地球村、公众环境研究中心等。中国政府鼓励社会各方面包括 NGO 领域的专家、部门共同参与环境决策，并借助媒体、网络和通信等多种手段，建立面向公众的互动渠道。例如，为加强规划编制的科学性，提高规划实施效果，环境保护部组织的重点流域水污染防治等方面的规划编制过程中均强调了公众参与，采用了包括网络投票、咨询会议和规划信息公开等手段，强化了利益相关方的参与，并对 NGO 在规划实施中提出的监督性意见进行了积极反馈。

四、结论和展望

工业革命和信息革命以来,"绿色革命"已成为新一轮全球经济转型的催化剂。特别是自2008年绵延至今的全球经济危机,使人们更加清醒地认识到以资源耗竭与依赖化石燃料为特征的"黑色"经济发展模式是不可持续的,人类必须寻求一条绿色发展之路。作为一个新兴经济大国,发展绿色经济将对中国未来经济繁荣乃至全球经济发展产生深远影响。中国政府深刻认识到,中国资源环境面临的压力可能比世界上任何国家都大,环境资源问题比任何国家都要突出,解决起来也比任何国家都要困难。转变经济发展方式,实现绿色发展,这不仅是中华民族长远发展的战略性选择和必然需要,也是对全球可持续发展的积极贡献,将对人类发展产生重要影响。

"十一五"以来,中国的环境保护从认识到实践上均发生了重大转变。环境保护不仅被视为优化经济增长,倒逼经济增长方式向绿色转型的有效手段,而且也是提升中国经济竞争优势,抢占未来全球经济制高点的重要抓手。这一点已经成为各级政府和公众的共识。而且,通过"十一五"规划实施,中国政府全面完成了规划提出的各项约束性目标指标,并在该过程中,全面加强了政府环境监管能力,推动了环保产业和环保技术发展,这将为"十二五"乃至更长期的经济增长提供基础。

"十一五"中国环境政策领域的成功依赖下述三个要素:一是通过完善体制机制,建立了环境规划和环境政策的强有力的实施机制。从中央到地方,通过目标责任制和行之有效的绩效考核机制,确保了包括总量减排在内的各项环境指标的实现。同时,在环境领域,通过组建环境保护部,成立各大区域督查中心,强化了中央政府对环境规划和政策的实施监督。二是实施了一系列行之有效的政策手段。在节能减排领域,不仅强调通过法律和行政手段来强化政策目标对各级政府和企业的约束,同时也注重经济手段,如脱硫电价补贴政策的实施,对于鼓励电力企业自主减排起到了积极的推动作用。三是通过实施一系列能力建设,为各项政策措施的落实提供了基本保障。

中国仍将面临持续的资源环境压力。尽管"十一五"末以来,受全球金融和经济危机影响,中国经济出现下滑态势,但中国经济增长的趋势并没有改变,对能源资源的需求仍将保持快速上升态势,经济结构和增长方式转型仍将面临较大困难。同时,由于历史欠账较多，中国的生态环境状况总体恶化的趋势尚未得到根本遏制，环境风险仍处于高发状态。解决当前面临的复杂和复合型资源环境问题，不仅有赖于加快经济增长方式和结构调整步伐，更需要充分动员包括国际社会在内的力量，参与和支持中国的绿色发展进程。

"十二五"时期，中国确立了面向绿色发展的更具包容性的发展目标。中国"十二五"规划的总体目标是加快经济增长方式和结构调整，以实现包容、绿色以及有竞争力的经济发展模式，规划共包含了8个与绿色经济发展直接相关的宏观经济与环境发展指标。在统筹人口、资源与环境协调发展上，中国政府提出要实施区域发展总体战略和主体功能区战略，构筑区域经济优势互补、主体功能定位清晰、国土空间高效利用、人与自然和谐相处的区域发展格局，逐步实现不同区域基本公共服务均等化，使发展成果更多惠及公众。

附件 OECD中国环境绩效评估51条意见和建议落实情况

1 意见、建议

在全国范围内实施针对产品和工业/能源设施的环境法律法规;加强监测、监督和执法能力,包括将环境执法职能从地方环保局中独立出来。

2 意见、建议

考虑将国家环保总局升格为环境部;加强国家环保总局对地方环保局的监督管理能力。

落实情况

目前,中国在能源和资源管理、环境保护方面的法律法规总计有50多部,主要包括《中华人民共和国宪法》、《中华人民共和国环境保护法》、《中华人民共和国水法》、《中华人民共和国水土保持法》、《中华人民共和国电力法》、《中华人民共和国煤炭法》、《中华人民共和国节约能源法》、《中华人民共和国可再生能源法》、《中华人民共和国矿产资源法》、《中华人民共和国水污染防治法》、《中华人民共和国大气污染防治法》、《中华人民共和国清洁生产促进法》、《中华人民共和国循环经济促进法》、《中华人民共和国固体废物污染防治法》等。

"十一五"以来,针对工业污染和能源领域,中国环保法规和相关标准不断完善,修订并发布了《中华人民共和国可再生能源法修正案》,制定实施《中华人民共和国循环经济促进法》,相继出台《规划环境影响评价条例》、《废弃电器电子产品回收处理管理条例》等7项环境保护行政法规;同时,完成了60余项重点行业污染排放标准的修订,开展了1 050项国家环保标准制修订的工作,目前国家环保标准达1 300项,比"十五"期间增加502项。这些立法和标准为加强工业领域的污染防治,降低单位产品污染排放,提高能源使用效率提供了基本依据和保障。

"十一五"期间,以落实节能减排目标为基础,中国全方位加强了环境监管能力建设,编制实施了《国家环境监管能力建设"十一五"规划》,目前,已累计完成投资300亿元。通过五年努力,环境监管能力和水平大幅提升。

机构建设方面,2008年3月,环境保护部正式成立,先后增设污染物排放总量控制司、环境监测司、宣传教育司、核设施安全监管司、核电安全监管司、辐射源安全监管司和环境保护部卫星环境应用中心等机构。新成立的环境保护部加大了总量减排、环境监管、核与辐射监管等重大问题的统筹协调力度，以环境保护部的成立为契机，各省先后将省环保局升格为环保厅。

2006年以来，原国家环保总局先后成立华北、华东、华南、西北、西南、东北6个区域环境保护督查中心；1999年以来，先后成立北方、上海、广东、四川、东北、西北6个区域核与辐射安全监督站。其中各环境保护督查中心为环境保护部的派出机构，其主要职能是监督地方对国家环境政策、规划、法规、标准执行情况；承办重大环境污染与生态破坏案件的查办工作；承办跨省区域、流域、海域重大环境纠纷的协调处理工作。部分省市还成立独立的区域督查中心，例如：江苏分别设立苏南、苏中、苏北三个区域环境保护督查中心、陕西成立陕北环境保护督查中心。
通过机构调整，进一步理顺了环境保护管理体制，基本确立了"国家监察、地方监管、单位负责"的环境监管体制。

3 意见、建议

继续努力促使地方领导对上级政府和当地群众承担更多的环境责任。

落实情况

政府绩效考核方面，2008年以来，监察部组织开展政府绩效考核工作，截至目前，全国共有24个省（区、市）和20多个国务院部门不同程度地探索开展了政府绩效管理工作。政府绩效考核工作以重点工作任务完成情况以及社会关注热点问题处理情况等作为考评重点，细化、实化、量化管理指标，使职责任务更加明确，过程控制更加有效，充分发挥了绩效管理的导向和激励约束作用。同时，还开展了一些针对地方政府的环境绩效考核工作。例如：通过"城市环境综合整治定量考核"和"创建国家环境保护模范城市"这两项环境管理制度，逐步建立起"地方政府领导，各部门分工负责，环境保护部门统一监督管理，公众积极参与"的城市环境管理工作机制，在督促地方政府加大环保投入、促进可持续发展等方面发挥了积极作用。

明确地方政府的环境保护和监管责任。现行《中华人民共和国环境保护法》明确规定："地方各级人民政府，应当对本辖区的环境质量负责，采取措施改善环境质量。"2008年2月，全国人大重新修订发布《中华人民共和国水污染防治法》，首次明确了将水环境保护目标完成情况作为地方人民政府及其负责人考核评价的内容，强化了地方政府水污染防治责任。另外，《中华人民共和国环境保护法》修改已列入十一届全国人大常委会立法规划，强化地方政府环境质量责任是本次修改的重点之一。各级政府除对环境质量改善负责外，还对环境风险防范、主要污染物总量减排承担责任。

总量减排作为各级政府承担的重要环境责任之一，中国政府建立了严格的监督制度。围绕总量减排目标，中国政府建立了目标分解、考核及奖惩机制。环境保护部代表国务院将减排目标分配到全国各省、自治区、直辖市和新疆生产建设兵团以及中国石油天然气集团公司、中国石油化工集团公司、国家电网公司、中国华能集团公司、中国大唐集团公司、中国华电集团公司、中国国电集团公司、中国电力投资集团公司 8 家中央企业集团，通过签订目标责任书的方式将减排目标分解落实。各省级政府又进一步将目标向下分解，从上至下形成了目标分解和考核机制。对于总量减排目标落实不到位的单位，严格考核机制，"十一五"期间，中央政府合计对 111 个地方政府或企业进行处罚，6 个地级政府、4 个集团公司被区域限批，26 位市长被约谈，100 多人被行政处罚。同时，全社会的污染治理能力、政府环境监管能力大大增强，环境质量不断改善。"十一五"期间污水处理厂累计增加约 2 000 座，建成投运的燃煤电厂脱硫设施增加 5.32 亿 kW，火电脱硫机组比例从 2005 年的 12% 提高到 2010 年的 82.6%；"十一五"期间 113 个环保重点城市的二氧化硫浓度下降 26.3%，759 个地表水国控监测断面 COD 浓度下降 31.9%，地级以上城市空气质量二级标准的比例明显提升。

环境监管方面，2011 年发布的《国务院关于加强环境保护重点工作的意见》中提出："要强化环境执法监管。执行流域、区域、行业限批和挂牌督办等督查制度。对未完成环保目标任务或发生重大突发环境事件负有责任的地方政府领导进行约谈，落实整改措施。"例如，对重特大突发环境事件频发、环境风险隐患突出的地区进行区域限批；2012 年环境保护部起草《突发环境事件调查与责任追究办法》，针对 2011 年以来环境保护部调度处置的重特大突发环境事件，对涉及事发地政府和有关部门不履行环境安全监管责任的，一律督促严肃追究责任。

为了促使地方政府承担更多的环境责任，资源消耗和环境保护作为考核的重要内容纳入了现行党政领导干部政绩考核内容。2006 年，中央组织部印发实施了《体现科学发展观要求的地方党政领导班子和领导干部综合考核评价试行办法》，进一步改进和完善干部考核评价工作，考核内容主要包括：本地人均生产总值及增长、人均财政收入及增长、城乡居民收入及增长、资源消耗与安全生产、基础教育、城镇就业、社会保障、耕地等资源保护、环境保护、科技投入与创新等方面。

4 意见、建议

强化综合的排污许可证管理，使其成为一个污染预防和控制的更加重要的手段；进一步将环境保护纳入土地利用规划和法规当中。

落实情况

环境保护部积极开展排污许可证的立法准备工作，努力推进立法进程。2007 年，原国家环保总局组织编制了《排污许可证管理条例（征求意见稿）》。2008年，新修订的《中华人民共和国水污染防治法》明确将排污许可证作为强化污染物排放监管的重要手段。从法律上确定了实施水污染物排放许可证制度的合法性地位，为实施排污许可证制度提供了法律依据。"十一五"以来，全国开展的主要污染物总量减排工作也为排污许可证的立法工作奠定了基础。另外，中国积极开展综合的排污许可证管理试点工作。自2004年以来，原国家环保总局就陆续在河北省唐山市、辽宁省沈阳市、浙江省杭州市、湖北省武汉市、广东省深圳市和宁夏回族自治区银川市开展排污许可证试点工作。

为综合考虑经济发展和环境保护问题，中国政府已经将环境保护纳入相关规划和法规之中。同时，2009年发布实施的《规划环境影响评价条例》（国务院令第559号）也提出："与土地利用的有关规划和区域、流域、海域的建设、开发利用规划，以及工业、农业、畜牧业、林业、能源、水利、交通、城市建设、旅游、自然资源开发的有关专项规划，应当进行环境影响评价。"中国国务院2008年印发的《全国土地利用总体规划纲要（2006—2020年）》，专门在第五章明确要求要"协调土地利用与生态建设"。《中华人民共和国土地管理法》修订已经列入十一届全国人大常委会立法规划，并由国务院组织起草。

国土空间开发方面，为了形成人口、经济、资源环境相协调的开发格局，中国政府2010年发布了《全国主体功能区规划》，提出将中国国土空间按开发方式分为优化开发区域、重点开发区域、限制开发区域和禁止开发区域；优化开发区重点是要加快转变经济发展方式，调整优化经济结构，提升参与全球分工与竞争的层次；重点开发区主要是在优化结构、提高效益、降低消耗、保护环境的基础上推动经济可持续发展，推进新型工业化进程，增强产业集聚能力；限制开发区主要是指农产品主产区，该区域重点是保护耕地，发展现代农业，增强农业综合生产能力，增加农民收入；禁止开发区是指重点生态功能区，主要是指在国土空间开发中限制大规模高强度工业化城镇化开发，以保持并提高生态产品供给能力。规划还要求调整完善财政、投资、产业、土地、农业、人口、环境等相关规划和政策法规，建立健全绩效考核评价体系。

同时，根据《国务院关于加强环境保护重点工作的意见》（国发[2011]35号），国家编制环境功能区划，在重要生态功能区、陆地和海洋生态环境敏感区、脆弱区等区域划定生态红线，对各类主体功能区分别制定相应的环境标准和环境政策。加强青藏高原生态屏障、黄土高原—川滇生态屏障、东北森林带、北方防沙带和南方丘陵山地带以及大江大河重要水系的生态环境保护。

5 意见、建议

扩大排污收费、使用者收费、排污交易和其他经济手段的使用范围,加强其激励作用,并将实施中的社会因素考虑进去。

落实情况

"十一五"以来,中国继续加快制定和实施一系列环境经济政策,各地在政策制定过程中也根据实际情况制定符合地方情况的环境经济政策,例如,《排污费征收使用管理条例》规定:国务院价格主管部门、财政部门、环境保护行政主管部门和经济贸易主管部门,根据污染治理产业化发展的需要、污染防治的要求和经济、技术条件以及排污者的承受能力,制定国家排污费征收标准。国家排污费征收标准中未作规定的,省、自治区、直辖市人民政府可以制定地方排污费征收标准。中国政府通过有效运用经济手段,促进节能减排和环境保护,推动产业结构调整,并取得积极成效。

排污收费制度规范化程度不断提高,覆盖面逐渐扩大,征收标准逐渐严格,征收稽查力度不断加大,排污收费总额不断提高,排污费应发挥的作用逐渐加强。例如,江苏、北京和天津等12个省份提高了二氧化硫排污收费标准。同时,为了应对中国污水处理收费标准偏低的问题,中国政府提出了城市供水设施污水处理"保本微利"的改革思路,2009年以来,在统计的36个大中城市中,已有10个城市较大幅度提高了污水处理收费标准,平均提高幅度29%,部分城市提高幅度近70%。排污费征管的信息化建设也逐渐受到重视,信息公开手段被应用于激励企业按时足额缴纳排污费,如对社会公布那些未按时缴纳排污费并予以罚款的企业名单。

排污权交易方面,2009年以来,环境保护部会同财政部,先后启动了浙江、陕西、湖南、湖北、山西、内蒙古、河北、重庆、河南等省市的排污权有偿使用和交易试点工作。其中,2011年浙江省印发了《浙江省排污权有偿使用和交易试点工作暂行办法实施细则》用于指导排污权交易工作。《国务院关于加强环境保护重点工作的意见》中明确提出"十二五"期间中国将建立国家排污权交易中心,发展排污权交易市场。目前中国部分省市已经成立了省级排污权交易中心,例如,浙江、北京、山西、重庆等。

环境税方面,中国目前没有专门针对环境保护的税收立法,并且相关部门正在积极开展环境保护税的研究工作。例如,中国政府颁布了《环境保护、节能节水项目企业所得税优惠目录(试行)》、《资源综合利用企业所得税优惠目录(2008年版)》、《节能节水专用设备企业所得税目录(2008年版)》、《环境保护专用设备企业所得税目录(2008年版)》、《关于调整完善资源综合利用产品及劳务增值税政策的通知》

等政策文件，对有利于环境保护的项目和设备通过所得税等税收政策予以扶持。另外，为建设"资源节约型、环境友好型"社会，促进节能减排，推动资源综合利用工作，经国务院批准，陆续出台了《财政部国家税务总局关于资源综合利用及其他产品增值税政策的通知》（财税[2008]156号）、《财政部 国家税务总局关于调整完善资源综合利用产品及劳务增值税政策的通知》（财税[2011]115号），对污水处理、垃圾处理等劳务以及再生水、以垃圾为燃料生产的电力或者热力等产品实行增值税优惠政策。因此，建议在该部分增加上述有关增值税优惠政策文件。

价格政策方面，中国实施的脱硫电价政策加快了火电行业脱硫设施的建设和运行，极大地促进了二氧化硫减排；同时，还加大了对高耗能企业差别电价政策实施力度，对超能耗产品实行惩罚性电价，差别电价和阶梯电价政策。"十一五"期间，实施了1.5分/度电的脱硫电价，中央财政补助40万元/km的污水处理厂管网配套建设补贴。同时，"十一五"期间，为了鼓励企业保护环境，大多数省份提高了污水处理费和垃圾处理费标准。为推动氮氧化物减排，2011年11月，在14个省份开展脱硝电价补贴试点。

绿色信贷方面，环境保护部、人民银行、银监会等部门近几年先后出台了一系列政策文件，如《关于落实环保政策法规防范信贷风险的意见》、《关于全面落实绿色信贷政策进一步完善信息共享工作的通知》、《关于进一步做好金融服务支持重点产业调整振兴和抑制部分行业产能过剩的指导意见》，明确要求金融机构对落后产能要严把信贷关。同时，《国务院关于加强环境保护重点工作的意见》提出健全环境污染责任保险制度，开展环境污染强制责任保险试点。

绿色贸易方面，取消了若干"高污染、高环境风险"产品的出口退税，并禁止其加工贸易。

上市公司和行业准入环保核查方面，2007年以来，环境保护部陆续发布了《关于进一步规范重污染行业生产经营公司申请上市或再融资环境保护核查工作的通知》、《关于加强上市公司环境保护监督管理工作的指导意见》、《关于进一步严格上市公司环保核查加强环保核查后督查工作的通知》和《关于进一步规范监督管理严格开展上市公司环保核查工作的通知》等文件，明确规定从事火力发电、钢铁等14个行业的公司应开展上市环保核查，通过环保后督查督促上市公司切实整改环保问题，持续改进环境行为；要求上市公司依法披露环境信息，促进公众参与和社会监督；通过上市环保核查引导资金投向资源节约、环境保护的行业和企业，促进经济发展方式转变。《国务院办公厅转发环境保护部等部门关于加强重金属污染防治工作指导意见的通知》和《国务院关于加强环境保护重点工作的意见》均明确要求，完善公司首次上市或再融资、资本重组环保核查制度，严格上市企业环保核查。

2001年以来，环境保护部持续开展了柠檬酸、味精、制革、稀土、钢铁、淀粉（淀粉糖）、酒精等行业环保核查。行业环保核查制度还与环保支持优惠政策挂钩，与相关部门联动，通过行业环保核查、符合环保要求的企业，在办理环评审批等行政许可、固体废物和危险化学品进出口审查、环保资金支持等方面都优先予以支持。

2008年7月，在国务院首次召开的全国农村环境保护工作电视电话会议上，李克强副总理提出了"以奖促治、以奖代补"等政策措施。中央财政首次设立农村环保专项资金，三年来共安排40亿元资金，支持了6 600多个村镇开展环境综合整治和生态示范建设，2 400多万农村人口直接受益。

6 意见、建议

实施更加有挑战性的大气污染削减目标，使其能够满足实现既定的大气环境质量目标的要求；加大污染物控制的范围，包括VOCs和有毒物质。

落实情况

为了进一步改善大气环境质量，针对污染特点、环境容量等情况，中国政府修订一些排放标准。例如，2012年，发布了新的《环境空气质量标准》，本次修订调整了环境空气功能区分类方案，修订了污染物项目及其限值，新标准中污染控制项目实现了与国际接轨，增加了细颗粒物（$PM_{2.5}$）浓度限值、臭氧（O_3）8h平均浓度限值等监测指标。2011年，环境保护部修订发布了《火电厂大气污染物排放标准》，调整了火电厂大气污染物排放浓度限值。

大气污染防治方面，中国政府建立了区域污染联防联控机制，在大气污染联防联控重点区域，建立区域空气环境质量评价体系，开展多种污染物协同控制。2010年，国务院办公厅发布了《关于推进大气污染联防联控工作改善区域空气质量指导意见》，"十二五"期间将进一步完善大气污染防治机制，以京津冀、长三角和珠三角等区域为重点，实施多污染物协同控制。

大气污染物总量减排方面，扩大了总量控制污染物种类，拓展了减排重点领域，强化了减排措施。《中国政府"十一五"国民经济和社会发展规划纲要》提出二氧化硫排放量降低10%的目标，建立了与之配套的目标责任制和配套的考核机制，经过五年努力，中国"十一五"污染减排目标全面完成，2010年二氧化硫排放总量2 185.1万t，与2005年相比，下降14.29%，超额完成10%的减排任务。《中国国民经济和社会发展第十二个五年规划纲要》又进一步提出了到2015年，单位国内生产总值能源消耗和二氧化碳排放分别降低16%、17%，二氧化硫减少8%，氮氧化物减少10%的总量控制目标。

污染控制范围不断扩大，规划提出"十二五"期间，加强挥发性有机污染物和有毒废气控制。2011年批复的《重金属污染综合防治"十二五"规划》中提出到2015年，重点区域铅、汞、铬、镉和类金属砷等重金属污染物的排放，比2007年削减15%的目标。北京、上海、广州等城市正在尝试将VOCs纳入常规监测范围。温家宝总理在2012年政府工作报告中明确提出，2012年底要在京津冀、长三角、珠三角等重点区域以及直辖市和省会城市开展$PM_{2.5}$等项目监测，2015年覆盖所有地级以上城市。2010年环境保护部联合其他部门共同发布的《关于加强二噁英污染防治的指导意见》，提出："在铁矿石烧结、电弧炉炼钢、再生有色金属生产、废弃物焚烧等重点行业全面推行削减和控制措施，到2015年，建立比较完善的二噁英污染防治体系和长效监管机制，重点行业二噁英排放强度降低10%，基本控制二噁英排放增长趋势。"为了保障控制目标的完成，环境保护部组织开展了一些专项调查，例如，消耗臭氧层物质专项调查、持久性有机污染物专项调查、汞污染排放源专项调查等。近年来，环境保护部联合多部门针对重金属、化学品环境管理和危险废物等有毒物质，组织开展了专项执法检查。

加强机动车污染防治。通过优化城市交通、实施严格的机动车排放标准、推行机动车环保标准、加速淘汰"黄标车"、提升燃油品质等综合措施，共同推进机动车氮氧化物减排。《关于国家机动车排放标准第四阶段限值实施日期的复函》要求，从2011年1月1日起，凡不满足国Ⅳ标准要求的车用气体燃料点燃式发动机与汽车不得销售和注册登记。同时，近年来，上海、北京、广州、深圳、南京等已提前应用了国Ⅳ标准。

7 意见、建议

进一步提高实施有效的大气环境管理所需要的监测数据的质量，并扩大监测范围（例如：污染源和污染物）。

落实情况

为了提高监测数据质量，新修订的《环境空气质量标准》对采样方法、数据统计有效性等方面做了更严格的规定，更新了监测分析方法标准。监测能力建设方面，2008年环境保护部增设了监测司，截至"十一五"末，全国环保系统已建立2 587个环境监测站，形成了国家、省、市、区县四级环境监测网络，国家空气质量监测网络由原来的113个环保重点城市扩大至338个地级以上的城市，全国共设1 436个空气质量监测站点，监测6种主要污染物，污染物监测种类新增加了臭氧、一氧化碳和$PM_{2.5}$。

环境保护部于2009年5月印发了《环境监测质量管理三年行动计划（2009—2011年）》，提出要积极推进环境监测质量制度建设，实施环境监测质量管理三年行动计划。2010年，组织开展了"第一届全国环境监测专业技术人员大比武"活动。2011年，举办了"全国环境应急监测演练活动"。

为保障环境监测数据质量，环境保护部开展了重点污染源自动监控能力建设。截至目前，共建成部、省、市三级污染源监控中心 349 个，监控重点污染源 15 559 家。同时，环境保护部先后制定了《污染源自动监控管理办法》、《自动监测数据有效性审核办法》、《污染源自动监控设施运行管理办法》和《污染源自动监控设施现场监督检查办法》等文件，提高环境管理数据的有效性。目前，实施自动监控的二氧化硫工业废气排放口 6 063 个、氮氧化物工业废气排放口 4 329 个，对工业废气污染源实施了有效监控。自动监测设备必须与环境保护主管部门直接联网，实时传输数据。对未安装自动监测设备或自动监测设备没有与环境保护主管部门联网的污染源，环境保护主管部门定期对其进行手工监测，其中国家重点监控企业的监测频次不少于每季度一次。

环境保护部通过加强国家重点监控企业自动监控、污染源监督性监测、环境监察执法、环境信息与统计四个方面的能力建设，正在努力建立一套科学、系统的主要污染物排放总量数据传输、核定、分析体系，一套污染源监督性监测和重点污染源自动在线监测相结合的环境监测体系，以及一套严格的、操作性强的污染物总量减排考核体系。

8 意见、建议

加强城市供水和污水处理设施的投资和管理以实现中国长远的目标（健康和水环境质量）；加强成本的回收（运行和投资费用）；提高污水处理厂的运行绩效；清晰地界定水务部门和地方政府的责任。

落实情况

2008 年 12 月，全国人大修订发布了《中华人民共和国水污染防治法》，明确县级以上地方人民政府应当采取防治水污染的对策和措施，对本行政区域的水环境质量负责，并首次明确了水环境状况信息由环保部门统一发布；2008 年，国务院"三定"规定进一步理顺了环保、水利等相关部门的职责分工，环保部门对水环境质量和水污染防治负责，水利部门负责水资源的调配，住房城建部门负责城镇供水、排水与污水处理设施的建设和管理，卫生部门负责饮用水的卫生监督监测。

中国政府自 2007 年 7 月起实施了修订后的国家《居民生活饮用水卫生标准》（GB 5749—2006），其中水质指标由原来的 35 项增加至 106 项，增加了 71 项；修订了 8 项。卫生部自 2007 年起推进全国城市饮用水卫生监测网络建设，目前国家饮用水卫生监测网已覆盖 31 个省份；地方卫生行政部门按照《中华人民共和国传染病防治法》和《生活饮用水卫生监督管理办法》，对饮用水供水单位实施卫生许可管理，严格查处违法行为。2012 年 1 月，印发了《卫生部关于加强饮用水卫生监督监测工作的指导意见》，对加强饮用水卫生监督工作，保障广大群众身体健康和生命安全，提出了一系列要求和措施。环境保护、水利等相关部门一直开展饮用水水源地的水质监测工作，2011年温家宝总理在政府工作报告中指出："十一五"期间，农业农村基础设施加快建设，完成7 356座大中型和重点小型水库除险加固，解决2.15亿农村人口饮水安全问题。

不断加强城市供水、排水和污水处理设施的投资和管理。2007年，财政部出台了《城镇污水处理设施配套管网以奖代补资金管理暂行办法》，决定在中央财政设立城镇污水处理设施配套管网建设专项奖励补助资金，采取以奖代补方式，用于支持城镇污水处理配套管网建设，鼓励提高城镇污水处理能力。2008年新修订的《中华人民共和国水污染防治法》第44条针对城镇污水处理做出明确规定："城镇污水处理设施的运营单位按照国家规定向排污者提供污水处理的有偿服务，收取污水处理费用，保证污水集中处理设施的正常运行。""十一五"期间全国累计新增污水处理厂2 000余座，新增污水处理能力超过6 500万t/d；截至2010年底，全国已建成城镇污水处理厂2 832座，总处理能力达1.25亿t/d，较"十五"期末翻一番。"十一五"期间，全国设市城市和县城公共供水日供水能力增加0.33亿m³，管网长度增加22.21万km，用水人口增加0.96亿人。截至2010年底，全国城镇（设市城市、县城和建制镇）供水人口6.3亿人，供水普及率达90.3%。

强化城市生活饮用水卫生监督管理和城镇污水处理设施运行监管。原建设部2004年发布了《关于加强城镇污水处理厂运行监管的意见》，住房和城乡建设部2011年还发布了《城镇污水处理厂运行、维护及安全技术规程》的行业标准。另外，环境保护部每年公布各地所有城镇污水处理厂名单，供社会公众监督，同时对污水处理厂主要污染物全年排放超标情况进行通报。在节能减排工作计划中，对污水处理厂按年度进行监管。"十一五"期间，城市污水处理率由52%提高到77%。同时，环境保护部将全部城镇生活污水处理厂纳入国家重点监控企业名单，要求全部安装在线监控系统并与环保部门联网；出台了《关于加强城镇污水处理厂污染减排核查核算工作的通知》（环办[2008]90号），会同有关部门出台了《城镇污水处理厂污泥处理处置及污染防治技术政策》（建城[2009]23号），对加强城镇污水处理厂运行监管、建设自动监控系统和中控系统、污泥处理处置等提出了明确的要求。从2007年起，环境保护部对城市污水处理厂建设滞后、污水处理费政策不落实、污水处理厂建成一年后实际处理水量达不到设计能力60%的、以及已建成污水处理设施但无故不运行的部分地区实行了区域限批。

9 意见、建议

继续加强工业部门的水污染控制，提高其用水效率；提高排污收费费率和取水费；确保污水处理厂的有效运行；将取水和排放许可与总量控制计划相衔接，保持河流的最小径流量和水质目标。

落实情况

《中华人民共和国水污染防治法》规定:"地方人民政府应当合理规划工业布局,要求造成水污染的企业进行技术改造,采取综合防治措施,提高水的重复利用率,减少废水和污染物排放量。对严重污染水环境的落后工艺和设备实行淘汰制度,禁止新建不符合国家产业政策的严重污染水环境的生产项目。"2008年以来,发展改革委、环境保护部会同工业和信息化部等部门先后在电镀、制浆造纸、发酵、制革等重点废水排放行业发布了一系列清洁生产评价指标体系;为加强铅蓄电池及再生铅行业管理,2011年发布了《关于加强铅蓄电池及再生铅行业污染防治的工作的通知》,要求新建涉铅的建设项目必须有明确的铅污染物排放总量来源。

工业用水费不断提高,进一步完善实施差别水价以及污水处理收费等相关政策。为提高用水效率,水利部门制定了行业用水定额,实施用水总量控制,并对火力发电、石油化工、钢铁等高用水工业进行节水技术改造,新型工业园区普遍发展和推广了循环用水和串联用水系统,积极推行废水"零排放"。2010年,工业和信息化部印发了《关于进一步加强工业节水工作的意见》,提出加快淘汰落后高用水工艺、设备和产品,大力推进节水工艺技术和设备。同时,中国政府在各个行业实施了阶梯水价制度,提高了用水标准。2006年,国务院发布的《取水许可和水资源费征收管理条例》提出,有关部门可适时合理调整水资源费征收标准,扩大征收范围,严格水资源费征收、使用和管理。针对生活部门,实施了阶梯水价制度,将用水总量与水价挂钩。全面落实污染者付费原则,完善城镇污水处理收费制度,2009年以来,国务院及有关部门相继出台了一系列相关法规和政策,逐步提高征收标准,满足城镇污水处理设施稳定运行需求。

《国家环境保护"十二五"规划》提出,加大重点地区、行业水污染物减排力度。在已富营养化的湖泊水库和东海、渤海等易发生赤潮的沿海地区实施总氮或总磷排放总量控制。推进造纸、印染和化工等行业化学需氧量和氨氮排放总量控制,削减比例较2010年不低于10%。严格控制长三角、珠三角等区域的造纸、印染、制革、农药、氮肥等行业新建单纯扩大产能项目。禁止在重点流域江河源头新建有色、造纸、印染、化工、制革等项目。

"十一五"以来,中国建立了总量减排目标责任制和相应的绩效考核机制,排污许可证作为落实总量减排目标的重要工具,在总量控制规划和年度计划的实施方面起到了积极作用。在主要污染物总量减排核查核算工作中,对企业取水、用水和排放情况进行了水平衡计算。2008年修订的《中华人民共和国水污染防治法》规定国家对重点水污染物排放实施总量控制制度和实行排污许可制度。环境保护部正在草拟的《排污许可证条例》将排污许可证与总量控制两项制度紧密衔接起来,许可证中将明确规定污染物排放的浓度标准和总量控制指标,对超过排放标准或者总量指标的依法予以处罚。为进一步强化排污许可证制度,《国家环境保护"十二五"规划》明确提出,要"全面推行排污许可证制度"。

10

意见、建议

继续努力加强农业水污染防治，提高用水效率，建立用水协会负责灌溉用水成本的回收；加强地下水开采费的监督和收缴；采取措施制止对地下水的过度开采；防止农业污染进入地下水含水层、河流和湖泊（例如：沿河和湖设立缓冲地带，规模化畜禽养殖污染的处理，有效地使用农用化学品）；废止对化肥的补贴。

落实情况

近年来，为了加强农业污染防治力度，中国政府在2007年开展了第一次全国污染源普查，初步摸清了农业污染源状况。《国务院办公厅转发环保总局等部门关于加强农村环境保护工作意见的通知》对全国农业和农村环境保护工作提出了具体要求。在农业面源污染防治方面，中央政府投入资金142亿元，对生猪、奶牛规模养殖场（小区）进行包括粪污处理设施在内的标准化改造，目前正在组织起草《畜禽污染防治条例》；测土配方施肥技术推广面积达11亿亩以上，在全国545个县（农场）实施土壤有机质提升补贴项目，严格限制高毒、高残留农药使用，已陆续淘汰33种农药，病虫害专业化系统防治和绿色防治实施面积达6.5亿。提高农田用水效率方面，中国大中型灌区实施了续建配套与节水改造，农田灌溉用水有效利用系数由2005年的0.45提高到2010年的0.50。2010年全国有效灌溉面积达到9.05亿亩，节水灌溉工程面积4.1亿亩。2005年，水利部、国家发改委、民政部联合印发《关于加强农村用水户协会建设的意见》，截至2011年底，全国共发展农民用水户协会7万多个，管理灌溉面积达2亿多亩。

2008年，中国政府设置了农村环保专项资金，通过"以奖促治"、"以奖代补"方式，鼓励农村环境综合整治和生态建设。2008年至2011年，共安排专项资金80亿元，带动地方投入近120亿元，支持1.7万个村镇开展环境综合整治和生态示范村镇建设，4 000多万农村人口直接受益。"十二五"对农业和农村环境保护提出更高要求。农业尤其是畜禽养殖业纳入减排范畴，同时，环境保护部和农业部开展了有关畜禽养殖污染防治和农业源减排等方面的合作。

中国政府十分重视地下水管理和保护工作，《国务院关于实行最严格水资源管理制度的意见》要求实行地下水取用水总量控制和水位控制，到2020年地下水超采得到根本遏制。为推动全国地下水污染防治工作深入开展，环境保护部会同有关部门编制发布了《全国地下水污染防治规划》。对于地下水的环境影响评价必须设置专章，对于规模化畜禽养殖场的选址提出明确要求，划定了禁养区。同时对于畜禽养殖粪便处理后安全排放提出了明确要求。国土资源部会同有关部门编制了《全国地面沉降防治规划（2011—2020年）》，该规划于2012年3月印发实施，明确了中国未来10年以控制地下水超采为主要措施的地面沉降防治目标和任务。水利部组织编制完成了全国地下水开发利用保护规划，划分地下水超采区，公布禁采和限采范围，编制南水北调东中线受水区地下水压采总体方案，采取措施治理地下水超采，江苏苏锡常地区、浙江杭嘉湖地区等地已全面禁采地下水。

11 意见、建议

进一步完善流域综合管理手段，改善水资源和水质管理，更加有效地提供与环境有关的服务（例如：洪水和干旱的预防，水土保持，生物多样性保护，为娱乐和旅游提供支持）；更加重视保护水生态系统（例如：河湖堤岸的复原，湿地的保护）；鼓励利益相关方的参与（例如：经济部门的代表，环境 NGO，专家，管理部门）。

落实情况

2002年修订通过的《中华人民共和国水法》确立了水中长期供求规划制度，强化了对流域综合规划的实施和监督。2008年6月1日起实施的《中华人民共和国水污染防治法》明确了地方政府的环境责任，为加强中国水污染防治提供了基础保障。

针对近年来水资源日益稀缺和水生态退化现状，中国政府提出让江河湖海休养生息、实施水资源和水环境综合管理的战略，强化了部门协调和任务分工。同时，针对中国水资源分布的区域特征，实施南水北调等重大水利工程，实施水资源空间配置均衡。同时，对大江大河水源地专门制定保护规划，严格限制各类开发活动，并对因水生态保护而受损的主体提供生态补偿。

为确保各项政策和措施的实施，中国政府建立了严格的目标分解和考核机制。尤其是在环境保护领域，中央政府建立了重点流域水污染防治专项规划实施情况考核制度，对地方政府省界断面水质考核情况和专项规划实施情况进行考核，有关结果向社会公布，进一步明确了地方政府的治污责任，有力地推动了流域水污染防治工作。同时，充分发挥区域环境督查派出机构的作用，加强区域流域环境问题的统筹协调；调整和完善了全国环境保护部际联席会议，建立了重点流域水污染防治部际联席会议，加强了部门、地方之间的沟通合作。另外，中央水利工作会议和2011年中央一号文件均要求实行最严格水资源管理制度，建立水功能区限制纳污制度。水利部为加强水功能区管理，推进水功能区限制纳污红线指标分解，开展全国重要江河湖泊水功能区纳污能力核定和分阶段限制排污总量控制方案制订工作。

为了提高水污染防治水平，"十一五"期间，中国实施了"水体污染控制与治理科技重大专项"，已启动32个项目，230个课题，中央财政经费32亿元已全部落实，攻克了一批难题，为水污染治理提供了技术支撑。

2008年，环境保护部发布了《淮河、海河、辽河、巢湖、滇池、黄河中上游等重点流域水污染防治规划（2006—2010年）》即水污染防治"十一五"专项规划，目标要使淮河、海河、辽河、巢湖、滇池、黄河中上游 6 个重点流域集中式饮用水水源地得到治理和保护，跨省界断面水环境质量明显改善，重点工业企业实现全面稳定达标排放，城镇污水处理水平显著提高，水污染物排放总量得到有效控制，流域水环境监管及水污染预警和应急处置能力显著增强。2011 年 3 月 16 日至 4 月 3 日，经环境保护部会同国务院相关部门对水污染防治"十一五"专项规划考核，重点流域水污染防治工作成效显著，水质达标率、规划项目完成率和投资完成率显著提高。

"十二五"期间，流域污染防治要在继续突出重点的同时，把覆盖范围扩大到所有大江大河大湖和有关海域，并实行分区控制，优先防控重点单元。通过财税优惠、项目倾斜等措施，鼓励一些地方率先摘掉流域水污染严重的帽子，让其休养生息。2010 年，辽宁省为加强辽河保护区管理，组织成立了"辽河保护区管理局"。在"十一五"规划的基础上，2012 年 5 月，环境保护部会同国务院有关部门印发《重点流域水污染防治规划（2011—2015 年）》，规划范围包括松花江、淮河、海河、辽河、黄河中上游、太湖、巢湖、滇池、三峡库区及其上游、丹江口库区及上游 10 个流域，目标到 2015 年，城镇集中式地表水饮用水水源地水质稳定达到功能要求；跨省界断面、污染严重的城市水体和支流水环境质量明显改善，重点湖泊富营养化程度有所减轻，水功能区达标率进一步提高；滇池湖体水生态系统明显改善；辽河流域率先由污染治理转入生态恢复阶段；主要水污染物排放总量和入河总量持续削减；水环境监测、预警与应急能力显著提高。

2011 年实施的《中华人民共和国水土保持法》提出："在水力侵蚀地区，应以天然沟壑及其两侧山坡地形成的小流域为单元，因地制宜地采取工程措施、植物措施和保护性耕作等措施，进行坡耕地和沟道水土流失综合治理"等各项预防和治理水土流失的政策措施。

中国十分重视湿地保护工作，"十一五"期间发布实施了《全国湿地保护工程实施规划（2005—2010 年）》，目前，初步建立了以湿地自然保护区、湿地公园为主的湿地保护网络体系，并建立了中央财政湿地保护补助专项，用于补助国际重要湿地、湿地自然保护区和国家湿地公园，开展湿地监控监测和生态恢复等工作，近 5 年，中国投入 30 多亿元恢复湿地近 8 万 hm^2。同时，中国认真履行《湿地公约》，强化中国履行《湿地公约》国家委员会的作用，顺利实施中澳、中德、中美等国际合作项目。

在水环境管理工作中，中国政府鼓励社会各方面包括 NGO 领域的专家、部门共同参与。例如，为加强规划编制的科学性，提高规划实施效果，环境保护部组织的重点流域水污染防治等方面的规划编制过程中均强调了公众参与，采用了包括网络投票、咨询会议和规划信息公开等手段，强化了利益相关方的参与。同时，政府对 NGO 在规划实施中提出的监督性意见进行积极反馈。

12 意见、建议

通过下列方法进一步鼓励水的可持续使用：①在制度上将水质和水投资结合起来（例如：在国家和其他相关层面的政府部门）；②在水服务进一步向全成本定价转变的过程中综合考虑市场的影响，对贫困地区和西部的特殊需要给予关注；③在水的立法和土地使用的改革中，明确和保证取水、分配和用水的权利。

落实情况

中国是一个水资源匮乏的国家，中国政府十分重视水资源管理工作。2011年，中国政府发布《中共中央、国务院关于加快水利改革发展的决定》（中发[2011]1号），提出要切实增强水利支撑保障能力，实现水资源可持续利用。

为进一步鼓励水的可持续使用，近几年，中国政府采取了多项措施。2006年2月21日，国务院发布了《取水许可和水资源费征收管理条例》，这对于加强国家对水资源的统一管理和监督，促进水资源的节约、保护和合理开发利用，建设资源节约型社会具有十分重要的意义。2008年4月9日，水利部发布了《取水许可管理办法》，加强了取水许可管理，规范取水的申请、审批和监督管理。2009年1月，水利部、财政部和国家发展改革委联合颁布了《水资源费征收使用管理办法》，31个省、自治区、直辖市全面征收水资源费，加大了征收管理力度。2011年，财政部、水利部联合印发了《中央分成水资源费使用管理暂行办法》，提高了中央分成水资源费使用效益。另外，2011年7月，财政部会同水利部印发了《关于从土地出让收益中计提农田水利建设资金有关事项的通知》，要求各地从2011年7月1日起，按规定口径从土地出让收益中计提10%农田水利建设资金，专项用于农田水利设施建设，重点支持粮食主产区、中西部地区和革命老区、民族地区、边疆地区、贫困地区农田水利建设。

中国《国民经济和社会发展第十二个五年规划纲要》提出要实行最严格的水资源管理制度，2012年国务院发布《关于实行最严格水资源管理制度的意见》（国发[2012]3号），提出要确立水资源开发利用和用水效率控制红线以及水功能区限制纳污红线等目标，严格实行用水总量控制、用水效率控制和入河湖排污总量控制。

为保障用水主体的权利，深入推进水权制度和水价改革，中国政府在全国各个层面开展了水权和水价制度改革试点。通过实施阶梯水价制度，强化水资源有偿使用和节约，通过对欠发达地区尤其是大江大河上游实施财政转移支付和生态补偿，补偿受损主体。一些省份出台了规范生态补偿的专门文件，如广东省2012年正式出台《广东省生态保护补偿办法》，将生态补偿分为基础性补偿与激励性补偿，基础性补偿将保证其基本公共服务支出需要，激励性补偿则与重点生态功能区保护和改善生态环境的成效挂钩，生态保护越好，获得奖励越多。

13 意见、建议

通过废物减量、废旧原料的再利用、废物循环和相关目标的制定来培育和发展循环经济;需要省级和地方政府实施全面的涵盖各种废物的综合管理计划。

落实情况

"十一五"以来,国家逐步加大了对于废物处理、资源再利用方面的政策制定和相关政策贯彻实施的力度。2007年3月,商务部、国家发展和改革委员会、公安部、原建设部、工商总局、原环保总局联合发布了《再生资源回收管理办法》。为加强对电子废物管理,原国家环保总局于2007年9月发布了《电子废物污染环境防治管理办法》。工业和信息化部已于近日公布了首批《再生资源综合利用先进适用技术目录》,涵盖废弃电器电子产品、废旧轮胎橡胶、废旧金属和废玻璃、废塑料和废纺织品、建筑和农林废弃物、废纸张及其他6大类产品综合利用产业领域。2009年1月1日起开始施行《中华人民共和国循环经济促进法》,促进了再生资源回收,规范了再生资源回收行业的发展。同时,自2006年商务部开展再生资源回收体系建设试点工作以来,截至目前共开展了三批90个城市试点。

"十二五"期间,中国制定了更高的固废管理目标。《国民经济和社会发展第十二个五年规划纲要》中提出:"工业固体废物综合利用率达到72%,资源产出率提高15%;健全资源循环利用回收体系。"为了贯彻《国务院关于印发"十二五"节能减排综合性工作方案的通知》的要求,工业和信息化部印发了聚氯乙烯、铜冶炼、铬盐等27个行业清洁生产技术推行方案,促进有关工业行业提升清洁生产技术水平;联合财政部组织实施清洁生产技术改造示范工程;2012年1月发布了《工业清洁生产推行"十二五"规划》,提出在"十二五"期间,工业清洁生产推进机制进一步健全,技术支撑能力显著提高,清洁生产服务体系更加完善,重点行业、省级以上工业园区企业清洁生产水平大幅提升;住房和城乡建设部于2012年4月发布了《关于加快推动我国绿色建筑发展的实施意见》,提出到2020年,建筑建造和使用过程的能源资源消耗水平接近或达到现阶段发达国家水平。

近年来,商务部一直积极开展再生资源回收体系建设工作。截至目前,共培育了90个试点城市,支持建设了91个区域性废旧商品回收利用基地。同时,商务部会同有关部门,积极开展再生资源回收行业的统计调查、立法和规划、标准制修订等各项基础工作。

为加快建立完整先进的废旧商品回收体系,2011年10月,国务院办公厅下发了《关于建立完整的先进的废旧商品回收体系的意见》(国办发[2011]49号)。2012年5月,国务院办公厅批准成立由商务部牵头、22个部门组成的废旧商品回收体系建设部际联席会议制度。2012年6月召开的部际联席会议第一次会议上，研究确定了废旧商品回收体系建设部际联席会议工作规则和2012年重点工作任务，将重点做好废旧商品回收行业税收、立法政策调研及统计等基础工作，开展废旧商品回收利用宣传。

对于固体废物综合管理方面，2010年，环境保护部、商务部、国家发展改革委、海关总署和国家质检总局颁发《固体废物进口管理办法》，环境保护部印发《进口可用作原料的固体废物环境保护管理规定》，严格可用作原料的固体废物进口技术审查和环境管理，充分发挥环境保护促进经济发展方式转变的综合作用，协同促进产业结构调整和污染减排，规范提升再生资源行业污染防治水平。2009年，环境保护部发布了《综合类生态工业园区标准（试行）》，该标准要求园区内必须具备废水、固体废物（包含废旧电子产品等）收集系统和废水集中处理设施。2010年，国家发展改革委和财政部发布了《关于开展城市矿产示范基地建设的通知》，计划通过5年的努力，在全国建成30个左右"城市矿产"示范基地。重点推动报废机电设备、电线电缆、家电、汽车、手机、铅酸电池、塑料、橡胶等重点"城市矿产"资源的循环利用、规模利用和高值利用。工业和信息化部积极推动工业循环经济示范工程建设，在广西河池、河北承德等12个地区推进大宗工业固体废物综合利用工作，提高综合利用水平，发布了《大宗工业固体废物综合利用"十二五"规划》，提出到2015年，大宗工业固体废物综合利用量达到16亿t，综合利用率达到50%，年产值5 000亿元，提供就业岗位250万个。

2009年以来，商务部会同财政部已支持全国100多家报废汽车回收拆解企业进行升级改造示范工程试点，并对相关示范工程给予最高为项目总投资额50%的资金支持，以提升报废汽车回收拆解行业的经营管理、资源综合利用和环境保护水平。

14 意见、建议

通过建设废物处理设施和建立废物的收集、再利用和再循环系统（例如：家庭废物分类收集）来加速提高废物处理能力，包括在农村地区。

落实情况

2006年以来，中国发布了一系列文件来推进废物回收利用工作。目前，商务部已经在全国90个城市先后开展了再生资源回收体系建设试点；2008年8月20日，国务院发布了《废弃电器电子产品回收处理管理条例》，以规范废弃电器电子产品的回收处理工作；2009年，商务部、财政部进一步提出加快推进再生资源回收体系建设。住房和城乡建设部也组织开展了对大中城市生活垃圾进行分类回收。2012年4月，为深入贯彻落实《国务院办公厅关于建立完整的先进的废旧商品回收体系的意见》，中华全国供销合作总社发布了《关于加快推进供销合作社废旧商品回收利用体系建设的意见》，推进供销合作社废旧商品回收利用体系建设，提出到"十二五"末，全系统废旧商品回收总额占全社会回收总额的比重达到 60% 以上的目标。

近几年，中国针对专门领域开展了具体的废物回收工作。例如，2009 年 6 月，国务院办公厅印发了《关于转发发改委等部门促进扩大内需鼓励汽车家电以旧换新实施方案的通知》，在北京、天津、上海、江苏、浙江、山东、广东、福州和长沙等省市开展家电"以旧换新"试点工作。财政部制定了《家电以旧换新运费补贴办法》。家电以旧换新政策的实施，对促进资源有效利用发挥了积极作用。2011 年，环境保护部印发《关于开展废铅酸蓄电池经营单位专项检查工作的通知》（环办函 [2011]470 号），加强持危险废物经营许可证的废铅酸蓄电池利用处置单位的环境管理，防止重金属污染。

针对生活垃圾处理方面，2007 年，原建设部发布了《城市生活垃圾管理办法》，对生活垃圾的清扫、收集、运输和处理提出了具体要求。2012 年 4 月，国务院办公厅印发《"十二五"全国城镇生活垃圾无害化处理设施建设规划》，提出到 2015 年，直辖市、省会城市和计划单列市生活垃圾全部实现无害化处理，设市城市生活垃圾无害化处理率达到 90% 以上，县县具备垃圾无害化处理能力，县城生活垃圾无害化处理率达到 70% 以上，全国城镇新增生活垃圾无害化处理设施能力 58 万 t/d。

农村生活垃圾处理方面，2007 年起，农业部组织在全国农村地区开展农村清洁工程示范建设，实施农业生产和农民生活废弃物的收集、处理和循环利用，示范村作物秸秆、人畜粪便、生活垃圾和污水等废弃物的资源化利用率达 90%。

15 意见、建议

针对不同部门（住户、大型企业、中小企业）和各种废物制订强制性计划。

落实情况

目前，中国主要对危险废物、医疗废物、放射性废物、危险化学品等废物实行强制性计划：

危险废物方面，中国制定了《中华人民共和国固体废物污染环境防治法》和《危险废物经营许可证管理办法》，对从事收集、贮存、利用、处置危险废物经营活动的单位实行危险废物经营许可证制度。产生危险废物的单位必须按照国家有关规定制订危险废物管理计划。2011 年，环境保护部、卫生部联合出台《关于进一步加强危险废物和医疗废物监管工作的意见》，对危险废物和医疗废物提出了更为严格的监管要求。2008 年，环境保护部启动修改《危险废物转移联单管理办法》，继续开展危险废物焚烧单位专项检查，加强对产废单位的管理。2007 年 9 月，原国家环境保护总局发布了《电子废物污染环境防治管理办法》，以防治电子废物污染环境，加强对电子废物的管理。为确保危险废物得到妥善处置，截至 2011 年年底，中国已有约 1 500 家持危险废物经营许可证的单位从事危险废物的收集、储存、利用、处置经营活动。

医疗废物管理方面，国务院于 2003 年公布实行《医疗废物管理条例》，推行医疗废物集中无害化处置，实行医疗废物分类收集、运送、储存、处置全过程管理。依据该条例，卫生部制定了《医疗卫生机构医疗废物管理办法》，并与原国家环保总局联合制定了《医疗废物分类目录》、《医疗废物管理行政处罚办法(试行)》、《医疗废物专用包装物、容器标准和警示标识规定》等一系列部门规章、规范及标准，以加强医疗废物的规范化管理；为进一步加大医疗废物监督管理力度，印发了《卫生部办公厅关于加强医疗卫生机构医疗废物监督管理的通知》，对全面加强医疗废物管理工作提出了具体措施。同时，卫生部与环境保护部多次联合组织对各地医疗机构、医疗废物集中处置单位的医疗废物规范化管理和无害化处置工作进行检查。

放射性废物管理方面，从 1994 年开始，中国就实施了《放射性废物管理规定》，明确了对放射性废物的产生、收集、处理、运输、储存及处置等各个环节在设计和运行中的管理目标和基本要求。针对放射性废物的安全管理工作，2011 年 12 月，中国政府又发布了《放射性废物安全管理条例》（国务院令第 612 号），该条例对放射性废物的处理、储存和处置及其监督管理等活动，提出了要求。

危险化学品管理方面，中国对危险化学品实行登记制度。2012 年修订的《危险化学品登记管理办法》，对登记内容、登记程序、登记企业的有关职责以及监督管理要求均做出了规定。目前，为了减少危险化学品向环境的释放，防范环境风险，环境保护部正在研究制定《危险化学品环境管理登记办法》。

16

意见、建议

理顺不同废物的管理责任；确保废物处理设施有效运行并满足标准；进一步建立实用的废弃物管理的法规和政策机制；加强废物数据资料的收集，开发评估国家和省级废物管理政策有效性的工具。

落实情况

废物管理方面，根据《中华人民共和国环境保护法》和 2004 年修订颁布的《中华人民共和国固体废物污染环境防治法》，各级政府环保部门负责固体废物、危险废物的监督管理；住房和城乡建设部门负责生活垃圾处置设施的建设和管理工作；商务部负责再生资源的回收利用工作。

为了进一步加强固体废物和放射性废物管理，近年来，中国先后发布了《电子废物污染环境防治管理办法》、《国家危险废物名录》（2008 年修订）、《废弃电器电子产品回收处理管理条例》和《放射性废物安全管理条例》等相关法律法规。对于固体废物污染环境防治的监督管理工作，《中华人民共和国固体废物污染环境防治法》

提出："建设项目需要配套建设的固体废物污染环境防治设施,必须与主体工程同时设计、同时施工、同时投入使用。同时,环境保护行政主管部门和其他固体废物污染环境防治工作的监督管理部门,有权依据各自的职责对管辖范围内与固体废物污染环境防治有关的单位进行现场检查。"

截至 2008 年,中国绝大多数省份都建立了固体废物管理中心和危险废物处置中心,其中部分城市也建立了固体废物管理中心,例如,深圳市;所有地级市都建立了医疗废物处置中心,并相应出台了医疗废物的收费办法。

废物数据收集管理方面,中国建立了大中城市废物信息发布机制,每年定期向社会发布固体废物相关信息。2009 年,环境保护部开展全国固体废物管理信息系统建设,系统建成后将进一步提高废物数据资料的收集能力。为了有效评估废物管理效果,从 2010 年起环境保护部在全国范围内开展危险废物规范化考核工作,促进各地贯彻落实危险废物管理的相关政策法规。

17 意见、建议

建立融公共和私人于一体的融资机制,在相对落后的地区进一步加强废物服务的收费;提高废物收费的收费率,并将费率设定在与政府要求实现的循环经济目标相一致的水平上。

落实情况

危险废物方面,为贯彻落实《中华人民共和国固体废物污染环境防治法》和《医疗废物管理条例》,加大危险废物处置力度、促进危险废物处置产业化,国家发展改革委、原国家环保总局、卫生部、财政部、建设部于 2003 年联合下发《关于实行危险废物处置收费制度促进危险废物处置产业化的通知》,要求各地合理制定危险废物处置收费标准,其收费标准应按照补偿危险废物处置成本、合理赢利的原则核定。危险废物处置成本主要包括危险废物收集、运输、贮存和处置过程中发生的运输工具费、材料费、动力费、维修费、设施设备折旧费等。广东、山东等省均进行危险废物收费制度。目前,废物处置的市场化机制逐步建立,已经形成了国家投资、地方和私人投资相结合的多元化投资模式。

电子废物方面,为了配合 2011 年开始实施的《废弃电器电子产品回收处理管理条例》,2012 年发布了与之配套的《废弃电器电子产品处理基金的征收使用管理办法》,该条例规定:"按照电器电子产品生产者销售、进口电器电子产品的收货人或者其代理人进口的电器电子产品数量定额征收,同时对处理企业按照实际完成拆解处理的废弃电器电子产品数量给予定额补贴。补贴标准为:电视机 85 元 / 台、电冰箱 80 元 / 台、洗衣机 35 元 / 台、房间空调器 35 元 / 台、微型计算机 85 元 / 台。"

生活垃圾方面，中国政府一直积极推进生活垃圾处理、收费工作。近年来，中国城镇生活垃圾收运网络日趋完善，生活垃圾处理设施数量和能力快速增长。截至2010年年底，全国设市城市和县城生活垃圾年清运量2.21亿t，生活垃圾无害化处理率63.5%，其中设市城市77.9%。2007年，建设部发布了《城市生活垃圾管理办法》规定："单位和个人未按规定缴纳城市生活垃圾处理费的，对单位可处以应交城市生活垃圾处理费三倍以下且不超过3万元的罚款，对个人可处以应交城市生活垃圾处理费三倍以下且不超过1 000元的罚款。"北京市2012年发布的《北京市生活垃圾管理条例》规定对于未按要求进行垃圾分类、处理、回收的单位，均提出了经济惩罚措施。

总体上，本着"减量化、资源化、无害化"的处理原则，近年来出台了一系列政策措施，逐步提高废物回收、处理标准。在中、西部欠发达地区，加大政府部门对废物处理设施建设的投资引导。废物资源回收方面，中国政府则出台鼓励民间资本进入资源回收利用领域的优惠政策，引导民间资本进入。目前，中国已经出现了一批由民间投资运营的资源回收企业。

18 意见、建议

作为废物管理计划的一部分，为非正规部门（垃圾收集者）提供设备、组织援助和培训，使其在改善的卫生和环境条件下，持续进行废物收集和再循环。

落实情况

目前，中国广泛开展了农民回收工培训工作。如中国再生资源回收利用协会成立了专门机构负责农民回收工的培训工作，开展了"服务新农村，培训农民回收工"等一系列培训活动，辐射带动全国各地再生资源行业协会培训农民回收工8万多人，各级供销合作社和中华全国供销合作总社有关直属单位也积极开展废物回收从业人员的培训工作，通过专业培训，使废物回收从业者掌握了分辨废物的知识，提高了分选、拆解的技能，改变了无序回收、二次污染的现象。

中华全国供销合作总社提出到"十二五"末，在80%以上的城市社区设立规范的回收站点。创新经营和回收方式，广泛开展电话预约、在线回收、以旧换新等灵活多样的回收方式，推动回收企业与机关单位、生产加工企业建立稳定的再生资源回收关系，建立多元化、多渠道的便民回收体系，进一步畅通回收渠道，提升网络运营水平。商务部将进一步落实国家"十二五"规划纲要提出的完善再生资源回收体系，加快建设城市社区和乡村回收站点、分拣中心、集散市场"三位一体"的回收网络，推进再生资源规模化利用。建设80个网点布局合理、管理规范、回收方式多元、重点品种回收率高的废旧商品回收体系示范城市的工作目标，按照《国务院办公厅关于建立完整的先进的废旧商品回收体系的意见》要求，借助废旧商品回收体系建设部际联席会议制度工作平台，以建立完整的先进的废旧商品回收体系为目标，从健全工作机制、做好舆论宣传、突出重点任务、强化基础工作、搭建工作平台、做好节能示范等几方面入手，争取到"十二五"末，逐步形成以龙头企业为主导、以回收站点为基础、以分拣加工集聚区为核心、以信息平台为支撑的废旧商品回收体系。

同时，NGO 组织如自然之友、地球村等也在开展各种活动以提高城乡居民的废物收集能力。

19 意见、建议

提高公众、中小型企业和产业部门的废物管理和资源有效利用的意识。

落实情况

近几年，环境保护部和相关部门已针对固体废物管理开展了多种形式的宣传活动，以提高公众及相关单位的废物管理意识。2010 年，组织了《废弃电器电子产品回收利用管理条例》及相关政策宣传贯彻培训会；针对新颁布的《固体废物进口管理办法》和《进口可用作原料的固体废物环境保护管理规定》等固体废物管理政策法规组织多次培训，提高废物进口企业的环保和法律意识，促进废物资源的有效利用。

2012 年 6 月 1 日，中国再生资源回收利用协会电子废弃物回收处理分会携手行业百家企业发出倡议，号召大家共同规范回收、环保拆解、守法经营、共创绿色家园。

为加强企业废物管理意识，从 2005 年开始，诺基亚在中国和合作伙伴一起开展了"绿箱子环保计划"，至今已在中国大陆约 300 个城市的 700 多个服务网点设置了废旧手机和附件回收箱。截至 2010 年年底，累计回收 160 余 t 废弃手机及附件，由诺基亚委托专业的公司进行环保处理，从废弃产品中得到金属和塑料等再生原材料。

政府部门为了推动资源节约型、环境友好型社会建设，近年来，国家发展改革委每年组织"全国宣传周活动"，大力宣传能源资源国情，强化公众资源忧患意识和节约意识，倡导全社会进一步把节能、低碳、绿色理念转化为全民行动。另外，根据国务院办公厅《关于建立完整的先进的废旧商品回收体系的意见》关于深入开展宣传教育的相关要求以及全国废旧商品回收体系建设部际联席会议 2012 年重点工作安排，拟由商务部联合中宣部、发展改革委、教育部、工业和信息化部、环境保护部、住房和城乡建设部、广电总局、国管局等部门于 2012 年 11 月共同举办全国"废旧商品回收利用"宣传活动。

全国废旧商品回收利用宣传活动的目的是宣传中国促进废旧商品回收利用和流通领域节能减排的政策措施和发展导向；推广开展回收体系建设取得的成效经验；介绍中国废旧商品回收利用行业对建设"两型社会"、促进国民经济和社会发展的重要意义和贡献；开展合理利用废旧商品的普及教育和勤俭节约的品德教育等。主要内容包括"绿色回收进校园""绿色回收进社区""绿色回收进机关""绿色回收进商场"等。

20 意见、建议

完善和实施自然保护的法律,尤其是对自然保护区条例进行修订。

落实情况

中国正在加快生态保护的立法工作,《中华人民共和国自然保护区法》、《中华人民共和国土壤污染防治法》、《中华人民共和国转基因生物安全法》、《中华人民共和国生态保护法》等法律正在制定。自然保护区立法目前已列入全国人大常委会的立法规划。

环境保护部、农业部、国家林业局等部委认真履行《自然保护区条例》规定的职能,积极推进全国自然保护区建设和发展。环境保护部联合有关部门开展国家级自然保护区管理评估工作,从2008年启动以来,已对22个省、自治区、直辖市的231处国家级自然保护区进行了管理评估。截至2011年年底,全国共建立自然保护区2 640处,保护区总面积约149万 km^2,陆地自然保护区面积约占国土面积的14.9%。

2010年12月,国务院办公厅印发《关于做好自然保护区管理有关工作的通知》,要求科学规划自然保护区发展,强化对自然保护区范围和功能区调整的管理,严格限制涉及自然保护区的开发建设活动,加强涉及自然保护区开发建设项目管理,规范自然保护区内土地和海域管理,强化对自然保护区的监督检查。通知同时还强调今后要加强领导和协调。

自然保护区建设正在实现由数量规模型向质量效益型的转变。同时,环境保护部正在启动实施全国生态环境十年(2000—2010年)变化遥感调查与评估工作,发布了《全国生态功能区划》,使区域生态功能保护水平进一步提升。另外,中国政府积极选划建设海洋自然保护区和特别保护区,提升海洋保护区规范化建设水平,严格监管海洋保护区周边开发利用活动,有效保护海洋保护区及邻近海域生态环境及生物多样性。

21 意见、建议

提高国家、省、市和县级机构实施生物多样性保护的能力,在实施保护区外的经济和社会发展项目时综合考虑自然保护的因素。

落实情况

中国是生物多样性公约的签约国,已于 1992 年 6 月 11 日签署该公约。1994 年 6 月,经国务院环境保护委员会同意,原国家环保总局会同相关部门发布了《中国生物多样性保护行动计划》。目前,该行动计划确定的七大目标已基本实现,26 项优先行动大部分已完成,随后,中国政府又先后发布了《中国自然保护区发展规划纲要(1996—2010 年)》、《全国生态环境建设规划》、《全国生态环境保护纲要》、《全国生物物种资源保护与利用规划纲要(2006—2020 年)》和《中国水生生物资源养护行动纲要》。相关行业主管部门也分别在自然保护区、湿地、野生动植物、水生生物、畜禽遗传资源保护等领域发布实施了一系列规划和计划。

环境保护部会同 20 多个部门和单位编制了《中国生物多样性保护战略与行动计划(2011—2030 年)》,该计划于 2010 年 9 月印发实施,明确了中国未来 20 年生物多样性保护总体目标、战略任务和优先行动。2011 年 6 月,国务院批准成立了"中国生物多样性保护国家委员会",由李克强副总理亲自担任主席,25 个成员单位的分管领导担任成员。委员会建立了长效工作机制,统筹协调全国生物多样性保护工作。环境保护部还积极推进生物多样性保护"主流化",加强宣传,促进各部门和地方将生物多样性保护纳入部门和地方有关规划和行动计划。

另外,中国开展"2010 国际生物多样性年中国行动"和"联合国生物多样性十年中国行动"并取得了明显成效,2010 年,全国共开展各类大型宣传活动 230 余项,影响受众达 9 亿多人次。

自然保护已经成为中国在保护区外实施经济和社会发展项目时考虑的重要因素。《开发区区域环境影响评价技术导则》中就明确规定,在对开发区进行环境影响评价时,应分析开发区规划实施对生态环境的影响。

22

意见、建议

增加自然和生物多样性保护机构的财政投入和人力资源,结合脱贫,让当地居民进一步参与到巡视、监督和栖息地增建计划中。

落实情况

多年来,各级政府和有关部门采取积极措施,按照《自然保护区条例》规定,将自然保护区建设管理纳入国民经济和社会发展规划,不断增加自然保护区的投入。1998 年,财政部设立了国家级自然保护区能力建设专项资金和林业国家级自然保护区建设补助资金,支持全国各类自然保护区建设,增强生物多样性保护能力。2012 年,财政部安排上述两项资金 4.3 亿元。2001 年,国家发展改革委批准实施了"全国野生动植物保护及自然保护区建设工程",对今后 50 年的全国野生动植物及自然保护区建设进行了全面规划和工程建设安排。

"十一五"期间，国家林业局全面实施以生态建设为主的林业发展战略，加速推进传统林业向现代林业转变，着力构建林业生态和产业体系，实施工程带动，转变增长方式，大力提高林业发展的质量和效益，不断开发林业的多种功能。自2007年以来，环境保护部共增加了约46%的行政编制，农业部、林业局等涉及自然和生物多样性保护职能的部门有关机构人员力量也都得到了加强。

中国也采取措施促进社区居民参与到自然保护区的建设和管理。在自然保护区设立时要充分征求社区居民意见，涉及居民承包的土地，还要与承包人签订委托管理协议。同时，很多自然保护区建立了社区共管机制，社区居民参与保护和管理的程度不断提高。

23 意见、建议

开发、使用与自然和生物多样性保护有关的经济手段，使其不仅作为增加收入的方法，而且作为提供环境服务的回报。

落实情况

在中国，开发、使用自然生态与生物多样性并对其予以反哺保护的经济手段主要是生态补偿政策。2007年8月，原国家环保总局发布了《关于开展生态补偿试点工作的指导意见》，为全国开展生态补偿工作提供指南。目前中国生态补偿实施方式主要是采取对一些重大生态工程建设项目进行投资，实施退耕还林补贴、草原生态保护补助奖励、森林生态效益补偿、农村沼气建设国债补贴、小型农田水利和水土保持补助费政策等方式。其中，中央财政森林生态效益补偿面积已扩大到13.85亿亩，草原生态保护补助奖励机制覆盖了全国64%的天然草原，全国所有省（自治区、直辖市）均建立并实施了矿山环境治理恢复保证金制度。生态补偿政策的实施对维护中国生态安全，促进生态环境服务均等化起到了重要作用。

中国政府积极推进生态补偿试点以及立法工作，出台了《重点生态功能区转移支付办法》。从2008年起，中国政府开始对三江源、东江源等重点生态保护区实施财政转移支付制度。2010年，国家批准山东威海、江苏连云港等沿海城市为海洋生态补偿试点地区。2010年，中国政府建立了湿地生态保护补助机制，对90多个国际重要湿地、湿地类型自然保护区和国家湿地公园加大了生态保护力度。

24 意见、建议

将恢复、保持物种和保护区（包括对外来物种的管理）的长期规划与土地使用和流域管理规划、相关省、市、县的规划相衔接。

落实情况

恢复、保持物种和保护区已经成为中国政府制定土地使用、流域管理规划、相关省、市、县的规划时所考虑的重要方面。《中华人民共和国野生动物保护法》明确规定禁止破坏国家保护野生动物及其生存环境，要求对建设项目可能影响到受保护野生动物生境的，在环评时须征求野生动物主管部门的意见。2004年，国务院办公厅发布了《关于加强生物物种资源保护和管理的通知》，明确要求各级部门要制定生物物种资源保护利用规划。根据国务院办公厅通知精神，原国家环保总局积极组织相关部委研究制定全国生物物种资源保护利用规划，并于2007年正式发布《全国生物物种资源保护与利用规划纲要》，纲要确立了到2020年全国生物物种资源保护与利用的规划总体目标和阶段目标，确定了今后15年内的保护与利用的重点领域以及近期优先行动领域与优先项目。

各省、自治区、直辖市也建立了本省的生物多样性保护工作协调机制，成立专家委员会，编制省级生物多样性保护战略与行动计划。部分地方开展了生物多样性保护、恢复和减贫示范，将生物多样性保护与消除贫困结合起来。部分区域和流域还结合国家生物多样性保护战略编制了本区域的生物多样性保护行动计划，如《滇西北生物多样性保护行动计划》。

此外，为加强生物物种资源保护和管理，统一组织、协调中国生物物种资源的保护和管理工作，2003年，由原国家环保总局牵头，建立了17个国务院有关部委组成的生物物种资源保护部际联席会议。自部际联席会议建立以来，环境保护部一直积极组织和协调各部委生物物种资源保护和管理工作，截至2012年4月，已组织召开了六届生物物种资源保护部际联席会议，为中国生物物种资源保护与利用规划奠定了基础。2004年，农业部牵头成立了7个国务院有关部委组成的全国外来入侵生物防治协作小组，农业部成立了外来物种管理办公室和外来入侵生物预防与控制研究中心，为外来物种管理内容纳入相关规划和指导全国开展外来入侵生物防治打下了基础。

25

意见、建议

将保护栖息地和物种的经济和社会价值（例如：生态服务，发展旅游）与发展决策相结合，尤其是要将其作为环境影响评价的一部分。

落实情况

国务院 2009 年颁布《规划环境影响评价条例》，明确要求把"规划实施可能对相关区域、流域、海域生态系统产生的整体影响"作为规划环评分析、预测和评估的内容。环境保护部在制定相关规划环评和项目环评技术导则时，把生态保护作为其中的重要内容。在执行过程中，对重大的开发建设项目进行环评审批时，栖息地和物种保护是重要的审查内容。

为了保护栖息地和物种资源，2008 年，环境保护部联合其他部门印发《关于加强自然保护区调整管理工作的通知》。2010 年，国务院办公厅又印发《关于做好自然保护区管理有关工作的通知》，强化了对自然保护区范围和功能区调整的管理和涉及自然保护区开发建设项目管理，环境保护部组织编制的《国家级自然保护区范围和功能区调整申报材料编制规范》，从严控制国家级自然保护区调整，严格涉及国家级自然保护区建设项目的生态准入，进一步规范了自然保护区调整评审工作。各地严格管理那些涉及自然保护区的开发建设项目的审批过程，制定出台了地方级自然保护区开发建设项目管理办法及环境影响评价审批规定。

26

意见、建议

建立一个跨政府部门的小组来评估如何调整与环境相关的税收，使其更有利于政策目标的实现。

落实情况

根据中国国情和行政管理体制的特点，环境保护部门加大与相关职能部门的协调配合，建立跨政府部门的环保合作机制。目前，财政部、税务总局和环境保护部三个部门正在研究起草环境保护税法草案，拟选择防治任务繁重、技术标准成熟的税目开征环境保护税，并逐步扩大征收范围。国家出台了对符合规定的节能减排项目以及购买符合规定的节能节水、环境保护专用设备的企业给予企业所得税的优惠政策；完善了对符合环境保护要求的资源综合利用产品及劳务的增值税优惠政策，对脱硫副产品、利用垃圾焚烧发电等给予增值税优惠政策。对大量消耗资源、污染环境的消费品征收消费税。车船税充分考虑环保因素，对乘用车按排气量大小实行差别化税率，体现了对小排气量乘用车的鼓励。同时，对节约能源、使用新能源的车船给予了减征或免征的优惠政策。2009—2010年对购买1.6升及以下排量乘用车给予减征车辆购置税的优惠。

除上述环境保护相关税收外，中国还通过贸易环节的退税和关税政策调整，鼓励清洁产品出口，抑制高污染产品出口，促进国内的资源节约和环境保护。自2007年起，环境保护部已经发布包含了514种"双高"产品和重污染工艺、42种环境友好工艺和15种污染减排重点环保设备的环境保护综合目录，同时，财政部、国家税务总局和海关总署取消了近200个税则号项下出口产品的出口退税，并禁止其加工贸易。

27

意见、建议

通过全面实施污染者付费和使用者付费原则，加强环境融资，增加融资渠道，提高公共环境支出分配的有效性和效率。

落实情况

中国正在形成多元化的环保投入的融资渠道。"十一五"环保投资总额达到2.16万亿元，其中中央财政资金1 564亿元，仅占总投入的7.24%。积极引入市场机制，健全城镇污水处理收费制度，实施特许经营，引导民间资本参与城镇污水处理设施建设和运营，截至2011年，全国共建成投运污水处理厂3 000余座，其中采取BOT/BT/TOT等特许经营模式的占42%。

中国已经实施污染者付费原则。2003年颁布实施的《排污费征收使用管理条例》（国务院令第369号）规定，直接向环境排放污染物的排污者，应当缴纳排污费。为保障依法、全面、足额征收排污费，2007年12月起开始实施《排污费征收工作稽查办法》（原国家环保总局令第42号）。

中国目前正在积极推进排污费和污水处理费的改革，主要目的在于促进污染者付费和使用者付费征收标准的提高，以抑制与减少环境行为的外部性，促进征收的规范化。

28

意见、建议

加强制度建设，更好地将环境问题纳入经济和部门决策，可能的话，建立一个有关环境或可持续发展的领导小组；全面实施环境影响评价法中有关对部门发展计划开展环境影响评价的规定。

落实情况

中国已经认识到综合环境经济决策对于保护资源环境和优化经济增长的重要作用,包括国家发展改革委、环境保护部、国土资源部、水利部、农业部等各大部委之间已经建立了行之有效的工作机制,协调经济发展和资源环境保护之间的矛盾。"十一五"以来,资源环境目标已经成为国家经济社会发展规划的重要约束性指标。"十一五"期间化学需氧量和二氧化硫排放总量分别下降12.45%和14.29%,均超额完成10%的减排任务。总量削减目标倒逼结构调整,淘汰落后产能,包括关停小火电机组7 200万kW,淘汰落后炼铁产能12 172万t、炼钢产能6 969万t、水泥产能3.3亿t。正在修改中的《中华人民共和国环境保护法》草案也明确提出了"经济建设和社会发展与环境保护相协调"的理念。

2006年在党的十七大报告中提出的"三个转变"(需求结构上的转变,产业结构上的改变,要素投入上的转变)重要思想,在"十一五"及以后得到了很好的实施。2007年6月,国务院成立由温家宝总理担任组长的国家应对气候变化及节能减排工作领导小组,研究审议重大政策问题,协调解决工作中的重大问题。2008年,在原国务院环境保护领导小组的基础之上建立了环境保护部际联席会议制度对生物物种管理和跨区域流域管理等重大环境问题协调决策机制。各地方也建立了类似的部门协调机制,例如浙江省宁波市就针对环境执法建立了环保部门与公安部门相互协调的联动机制。

在2011年召开的第七次全国环境保护大会上,李克强副总理充分肯定了"十一五"环保工作取得的显著成绩,系统分析了当前环境保护中存在的突出问题和深层次矛盾,明确提出要坚持在发展中保护、在保护中发展,积极探索代价小、效益好、排放低、可持续的环境保护新道路,切实解决影响科学发展和损害群众健康的突出环境问题,全面推进我国环保事业新发展。

同时,中国环境影响评价法律法规中有关对部门发展计划开展环境影响评价的规定已经全面实施。2009年8月,国务院第559号令发布了《规划环境影响评价条例》,从当年10月1日起开始实施。目前,实施环境影响评价已经成为制订部门发展计划的重要环节。内蒙古、重庆、深圳、大连等地通过地方立法,强化了规划环境影响评价的实施力度。2011年以来,环境保护部先后与国家发展改革委、交通运输部等部门联合发文,明确了相关领域的规划环境影响评价管理要求。总量减排目标已经成为环评审批的前置条件,实施一票否决。河南等地已经出台了有关前置审批的管理办法。

29

意见、建议

通过科学的经济和社会分析手段,继续设立国家目标以实现重要的环境目的。

落实情况

通过科学的经济和社会分析手段，中国设立了严格的国家目标以实现环境与经济的协调发展。环境问题在各领域"十一五"规划中得到了越来越多的重视；《中华人民共和国国民经济和社会发展第十一个五年规划纲要》提出了"十一五"期间单位国内生产总值能耗降低20%，主要污染物二氧化硫和化学需氧量排放总量减少10%的约束性指标。

《中华人民共和国国民经济和社会发展第十二个五年规划纲要》提出到2015年，单位国内生产总值能源消耗和二氧化碳排放分别降低16%、17%，化学需氧量和二氧化硫分别减少8%，氨氮和氮氧化物分别减少10%。同时，中国政府对重金属、VOC、$PM_{2.5}$等与公众健康紧密相关的污染物也纳入了环境管理重点范畴。

为了保证规划目标的科学性和可操作性，在科学测算基础上，总量减排任务分解过程经过了"两上两下"，即环境保护部制定一份"十二五"减排的指南给各个省、自治区、直辖市和新疆生产建设兵团以及8大中央企业，然后地方根据这个指南上报该地"十二五"减排计划的初稿，此为"一上"；随后环境保护部组织专家进行审查，再发给各地称为"一下"；地方根据该稿再进行修改，此为"二上"；最后经过中央同意，最终确定下发"十二五"的减排计划，称为"二下"。

意见、建议

减少不能接受到良好环境服务（安全用水、基本的卫生条件，用电）的人口比例，进一步提高健康和生活标准，尤其是在欠发达地区；在考虑可承受能力的情况下，在发展战略（例如：针对中国较穷的中西部地区）中优先考虑水基础设施。

落实情况

2005年国务院批准实施了《2005—2006年农村饮水安全应急工程规划》。2006年国务院又批准实施了《全国农村饮水安全工程"十一五"规划》，2005—2011年全国实际完成总投资1 360亿元，共计解决2.76亿农村居民的饮水安全问题。2008年环境保护部组织开展了全国饮用水水源地基础环境调查及评估工作，重点查明全国城镇和典型乡镇饮用水水源地环境基础状况，建立并完善集中式饮用水水源地基础信息，科学评估全国饮用水水源地基础环境状况。

2012年3月，国家又通过了《全国农村饮水安全工程"十二五"规划》，规划要求在持续巩固已建工程成果基础上，进一步加快建设步伐，全面解决规划内农村人口饮水安全问题，使全国农村集中式供水人口比例提高到80%左右。

"十一五"以来，中国积极推进医药卫生体制改革和基本卫生保健制度建设；加强医疗服务监管，努力解决群众看病难、看病贵问题；城乡卫生面貌发生变化，人民健康水平进一步改善；"户户通电"工程实施不到两年已经解决97.4万无电户360万人的用电问题。2003年起开始试点的新型农村合作医疗制度在2010年基本覆盖全国，新农合的实施大幅降低了农民的看病费用。

财政部在"十一五"期间逐渐加大中央财政对中西部地区均衡性转移支付力度，逐步缩小中西部地区地方标准财政收支缺口，推进地区间基本公共服务均等化。节能环保、新能源、教育、人才、医疗、社会保障、扶贫开发等专项转移支付，重点向中西部地区倾斜。

《国家环境保护"十二五"规划》将环境基础设施公共服务工程列为环境保护重点工程之一，包括城镇生活污染、危险废物处理处置设施建设，城乡饮用水水源地安全保障等工程。

2011年，中央发布一号文件强调加快水利建设，力争水利投入机制上有新突破，今后10年全社会水利年平均投入能够比2010年高出一倍。实行最严格的水资源管理制度，确立水资源开发利用控制、用水效率控制、水功能区限制纳污三条红线。针对水利又好又快发展的深层次制约，强调水利体制机制上要有新突破。

31 意见、建议

建立并利用环境绩效指标、与环境相关的经济信息分析、环境核算工具，如物质流核算，改进环境信息工作；扩大环境信息的覆盖面（例如：面源污染、有毒物质、危险废物）；使公众更好地掌握环境信息。

落实情况

中国基本环境管理制度中强调了环境绩效管理的作用，减排约束性指标，其中"创建国家环境保护模范城市"和"城市环境综合质量定量考核"作为两大实施手段有效地改善了城市环境质量，加强环境基础设施建设，充分发挥省级环保部门的作用。将"城考"和"创模"制度全面推广到全国范围。通过实施"城考"和"创模"制度并将其纳入地方政府政绩考核中，有利于进一步调动地方政府的积极性，全面推进我国城市的环境保护工作。

2007年国务院布置开展了第一次全国污染源普查，全面了解中国环境状况，掌握了各类企事业单位与环境有关的基本信息，建立健全了各类重点污染源档案和各级污染源信息数据库，准确了解污染物的排放情况，为正确判断环境形势、科学制定环境保护政策和规划、切实改善环境质量奠定了基础。"十二五"环境统计制度在污染源普查基础上根据环境管理的需要，从统计框架、统计范围和统计方法等方面进行了改革，为信息发布提供了基础。

中国正在逐步建立并利用环境绩效指标、与环境相关的经济信息分析、环境核算工具，扩大了环境信息的覆盖面，使公众更好地掌握了环境信息。从 2004 年至今，环境保护部一直在开展绿色国民经济核算工作。目前，正在进行上市公司环境绩效评估和上市公司环境信息公开的相关研究。2008 年 8 月，新版《国家危险废物名录》正式实施；2008 年 5 月 1 日，原国家环保总局颁布实施《政府信息公开条例》及《环境信息公开办法》，进一步规范和促进了政府和企业环境信息公开。

环境保护部连续发布环境统计年报和环境质量公报，同时在环境保护部网站主页上适时公布污染源监测信息。各级环境保护部门也通过质量公报和媒体等手段，发布与环境质量相关的信息。另外《大中城市固体废物污染环境防治信息发布导则》规定，大、中城市人民政府环保行政主管部门，应于每年 6 月 5 日前发布本城市上一年度固体废物污染环境防治的信息。

意见、建议

加强和完善健康和环境方面的信息，建立国家健康—环境行动计划；实施费效比最佳的措施；完善企业污染排放和转移报告制度；建立向环境健康风险（例如：职业健康、接近污染设施的健康影响，儿童健康）暴露人口发布风险报告的能力。

落实情况

环境与健康方面的信息进一步完善。危险废物转移联单制度已经有效实施多年。为更好地落实保护人民群众健康的宗旨，2007 年 11 月，卫生部、原国家环保总局牵头联合十六个部委局制定并发布《国家环境与健康行动计划（2007—2015 年）》，该行动计划是推动中国环境与健康工作科学发展的第一个纲领性文件。行动计划发布以来，环境保护部和卫生部充分发挥政府组织领导作用，积极促进公众参与。在能力建设上，多途径加强管理和技术支撑能力；在科学研究上，加大经费投入，加快科研成果为环境管理提供技术支持的步伐；在推动地方工作上，重点提高地方管理人员思想意识，推动各地环保部门认真落实行动计划。2010 年，经国务院批准，全国爱卫会在全国范围内开展了城乡环境卫生整洁行动，全面推动污水处理、垃圾清运、河沟整治、农村饮用水安全工程、农村改厕、标准化菜市场建设工程等工作，环境卫生面貌明显改善，为有效防控传染病发生、流行发挥了积极作用。

2007 年开始的第一次全国污染源普查将重金属等污染物排放及其环境风险列为重点调查内容。2009 年，环境保护部发布了《关于加强有毒化学品进出口环境管理登记工作的通知》，进一步规范了有毒化学品进出口环境管理登记审批，强化对进出口有毒化学品流向的监督管理，以全面防控有毒化学品在生产、使用、储运、处置过程中的环境风险。2010 年，环境保护部发布实施了《新化学物质环境管理办法》，有效防范新化学物质对人体健康和生态环境的影响，杜绝 POPs 类物质的产生。此外，按照巴塞尔公约和《中华人民共和国固体废物污染环境防治法》，还对跨境转移的可用作原料的固体废物进行了法律约束。

目前中国允许限制进口废纸、废金属、废五金电器、废船和废塑料等十大类废物，并将与人民健康密切相关的环境问题作为工作重点，如 $PM_{2.5}$、重金属、废铅酸电池等。并且针对电子废物拆解重灾区如广东贵屿等开展的专项防治规划、整治行动。

33 意见、建议

进一步扩大环境教育，提高环境意识，尤其是在年轻人当中。

落实情况

为加强中小学环境教育，推动绿色学校转型升级，在 1996 年开始的原国家环保局联合中宣部、教育部共同发起全国绿色学校创建活动的基础上，从 2010 年开始，开展了环境友好型学校试点工作，并开展了环境教育基地建设工作。

每年以"六·五"世界环境日为核心，中国都会开展不同专题的环境教育宣传活动，包括展览、演出、制作环境纪录片、绿色中国年度人物评选、中华宝钢环保奖评选等活动。2009 年 6 月 5 日，环境保护部联合全国人大环资委、全国政协人资环委、科技部、教育部、共青团中央、全国妇联启动了"千名青年环境友好使者行动"活动，面向全国选拔 1 000 名"青年环境友好使者"，这些志愿者深入社区、机关、学校、企业、公园和广场等，对公众开展节能减排宣讲，通过以一传千，带动百万青年积极参与环境保护。千名青年环境友好使者代表受到李克强副总理的亲切接见，并于 2011 年 12 月赴德班参加联合国气候变化大会，向国际社会充分展示中国青年应对气候变化的思考和行动。

2011 年 4 月 22 日，环境保护部、中宣部、中央文明办、教育部、共青团中央、全国妇联等六部门联合印发《全国环境宣传教育行动纲要（2011—2015 年）》，其中要求建设全民参与的社会行动体系，包括开展环境友好型学校、中小学环境教育社会实践基地建设等工作。

34 意见、建议

继续加强与 NGO 和公众的合作来实现环境政策目标；加强与企业的合作和伙伴关系。

落实情况

环境保护部一直十分重视环保民间组织对环境保护事业的推动作用,并积极支持和发挥NGO组织的作用,中国近年来涌现了一大批在国内外具有影响的环保NGO组织,例如自然之友、地球村、公众环境研究中心等。

环境保护部采取各种方式,加强与环保NGO组织的交流与沟通。针对环保NGO组织,环境保护部确定了"积极扶持、加快发展"、"加强沟通、深化合作"等基本原则。2012年4月,环境保护部组织召开了环保社会组织工作座谈会,潘岳副部长出席,与来自中国政法大学污染受害者法律帮助中心、自然之友、公众环境研究中心等20余家环保社会组织就如何推动公众及环保社会组织有效参与环境保护进行了面对面交流。2012年6月,环境保护部组织国内环保NGO参加在巴西里约召开的联合国可持续发展大会,推动了中国环保民间组织的国际交流与合作,展示了中国民间环保的力量和形象。从中央到地方,借助媒体、网络和通信等多种手段,建立了面向公众的互动渠道。

环境保护部配合全国人大常委会法工委,建议在修订后的《民事诉讼法》中规定环境公益诉讼。目前草案已经提出"对污染环境、侵害众多消费者合法权益等损害社会公共利益的行为,有关机关、社会团体可以向人民法院提起诉讼"。

近年来,环境NGO组织在中国的环保事业中也发挥着越来越大的影响。例如"地球村"已经从简单的社区卫生维护和垃圾分类发展到现在对政府有一定影响力的社团,其主要活动包括:建立绿色社区、培育生态乡村等。

意见、建议

继续开展国际环境合作,提高使用如下资源和方法的有效性和效率:①国内资源,②国际支持机制(例如:世界银行的清洁发展基金,在蒙特利尔议定书下的多边基金,全球环境基金)。

落实情况

为了推动国际环境合作,中国政府建立了由李克强副总理任主席的"中国环境与发展国际合作委员会",成立以来,国合会先后组建了涉及环境与发展问题诸多领域的几十个政策研究项目,提出了多项重要的政策建议。同时,1997年成立的环境保护对外合作中心,形成了以履行环境国际公约、开展双、多边环保项目合作、加强全球环境政策研究为主,以环保国际咨询和服务为辅的业务格局,进一步推动了中国环境保护对外合作工作。

近年来,积极利用国际支持机制与国内资源相结合,开展节能、温室气体减排、消耗臭氧层物质淘汰等国际环境合作，成绩显著：

1. 2006年7月以来，继续与全球环境基金（GEF）开展深入务实的合作，成功推动70多个国别项目获得GEF批准，获得赠款承诺约4.7亿美元，覆盖气候变化、生物多样性、土地退化、国际水域和持久性有机污染物五大重点业务领域。上述赠款资金充分发挥了杠杆作用，带动了大量公共和私营部门投入，进一步加强了中国履行相关国际环境公约的能力，有力地支持了国内生态环境保护和节能减排工作，提高了公众的环境保护意识。在气候变化领域，积极利用GEF赠款实施了"节能房间空调器推进"、"可再生能源规模化发展"等项目，并利用GEF赠款先后开展了"中国节能促进"、"中国节能融资"、"中国公用事业能效提高"（CHUEE）等创新项目，成功将"合同能源管理"节能新机制引入中国，培育壮大了中国节能服务产业。

2. 充分利用世界银行清洁发展基金。环境保护部开发的清洁发展机制（CDM）项目，其买家是世界银行管理的清洁发展基金。项目实施与温室气体减排成效显著，截至2011年年底，环境保护部开发的CDM项目获得联合国签发的减排量1.37亿t二氧化碳当量，占中国所有CDM项目同期签发总量的约29%，世界签发总量的17%，除促进资金转让外，与世界银行的合作项目还促进了HFC23分解技术的引进，在控制和减少温室气体排放同时，也提高了企业的环境管理水平，实现了多赢。

3. 有效利用蒙特利尔议定书多边基金。蒙特利尔议定书多边基金由发达国家和工业转型国家捐款，主要用于帮助发展中国家履行ODS削减的义务。自1991年至今，多边基金累计已增资20多亿美元，在约140个国家实施了5 000多个项目，用于支持包括工业转换、技术援助、信息传播、培训以及能力建设等活动，其投资重点为直接削减ODS的工业转换项目。截至目前，中国累计获得蒙特利尔议定书多边基金赠款约9.37亿美元，共实施了400多个单个项目和25个行业计划，受益企业3 000多家，在淘汰消耗臭氧层物质、保护臭氧层进程中发挥了巨大作用。

4. 欧盟项目：以中欧生物多样性项目和中欧流域管理项目为龙头，充分利用欧盟资金，促进国内环保工作开展。中国-欧盟生物多样性项目欧盟赠款3 000万欧元，是欧盟在海外资助的最大规模生物多样性项目，也是中国资金规模最大、覆盖地域最广、参与机构人员最多的生物多样性国际合作项目，已于2011年顺利完成，项目为推动中国生物多样性保护和可持续利用发挥了重要的作用，成为中欧环境领域成功合作典范。中欧流域管理项目计划总投资为1.865亿欧元，其中欧盟赠款2 500万欧元，中方配套资金7 900万欧元，世界银行贷款8 250万欧元，国内实施机构为环境保护部和水利部。项目旨在同欧洲和世界其他地区进行经验交流的基础上，探索并实施同全球当前所面临的社会、经济和自然变化相适应的流域综合管理模式，开发具体的系统和程序，在长江和黄河流域试点，并在可能的情况下将有关经验在其他地区推广。

5. 其他国际合作项目：以中意环保合作为例，中意合作成立了中意环保合作项目管理办公室，共实施合作项目90多个，支持了北京奥运、上海世博、四川汶川和青海玉树震后建设等重要活动和中西部发展，同时也通过湖泊治理、大气监测、光伏发电等项目，带动了意大利技术和产品进入中国环保市场，推动了企业间交流合作和国内相关技术水平的提升，达到互利共赢。在2010年温家宝总理访问意大利期间，中意环保合作成果得到了两国总理的高度肯定。

36 意见、建议

提高监测、监督和执法能力，为履行国际承诺提供保障（例如：有关濒危物种贸易、林产品，危险废物、臭氧层耗损物质，以及海洋倾倒和渔业管理）。

落实情况

监测、监督和执法能力不断提高，为履行国际承诺提供了有效的保障。中央财政于2007年设立主要污染物减排专项资金，支持环境监测、监察、应急、信息、宣教等环境监管能力建设，"十一五"期间，全国累计投资300亿元，其中中央投资约120亿元，同时，环境保护部组建了华北、华东、华南、西北、西南、东北6个区域环境保护督查机构，增设了监测司和卫星环境应用中心，并为中国环境监测总站增加了90名事业编制，有效地加强了国家和地方的环境监管能力。

危险废物管理方面，自1992年8月20日《巴塞尔公约》对中国生效后，中国政府采取有效措施，严格履行《巴塞尔公约》。颁布了符合《巴塞尔公约》要求的《危险废物名录》；认真执行《中华人民共和国固体废物污染环境防治法》和有关法律法规，严格审批检查手续，并出台专项法规对跨境转移废物进行严格控制。2008年，原国家环保总局出台了《危险废物出口核准管理办法》。在2005年6月20日《关于在国际贸易中对某些危险化学品和农药采用事先知情同意程序的鹿特丹公约》正式对中国生效前，中国已经将公约涉及的化学品和部分农药列入《中国禁止或严格限制的有毒化学品名录》，自愿执行了危险化学品和农药的进出口事先知情程序。

中国是《关于持久性有机污染物的斯德哥尔摩公约》的正式缔约方，是2001年5月首批签署公约的国家之一，加入公约以来，中国开展了杀虫剂类持久性有机污染物的削减和替代示范、持久性有机污染物废物处置和二噁英类重点排放行业减排技术示范等30多个国际合作项目，颁布了30多项与POPs污染防治和履约相关的管理政策、标准和技术导则，实现了滴滴涕等9种杀虫剂类持久性有机污染物的全面淘汰，处置了约2万t POPs废物和受污染土壤，维护了人民群众身体健康与环境安全。

为贯彻落实《中华人民共和国固体废物污染环境防治法》关于"产生、收集、储存、运输、利用、处置危险废物的单位，应当制定意外事故的防范措施和应急预案"的规定，指导危险废物经营单位制定应急预案，有效应对意外事故，2007年原国家环保总局制定了《危险废物经营单位编制应急预案指南》。

多边和双边环境监测合作方面,中俄自 2006 年以来每年都进行乌苏里江、兴凯湖、额尔古纳河、黑龙江和绥芬河等跨界水体联合监测,并向俄方无偿援助了水质检测仪、气相色谱仪和活性炭。为加强海洋倾废管理,中国政府还先后颁布了《海洋倾废管理条例》等一系列法规。

37 意见、建议

提高政府对海外中国企业的监督并改善其环境绩效(例如:依照 OECD 跨国企业指南来实施)。

落实情况

随着全球经济的发展和环境问题日益引起关注,中国政府对"走出去"企业的环境行为日益重视,有关部门正在完善中国企业境外环境行为指南,引导企业积极履行环境保护社会责任,保障中国对外投资的健康发展,促进投资所在地可持续发展。中国五矿集团公司自 2007 年每年发布《中国五矿集团可持续发展报告》,并参照"联合国全球契约十项原则"对集团履行情况进行考核。

在此之前,中国政府就已经认识到海外业务对环境政策的需求并有所行动。国家林业局会同商务部于 2007 年 8 月发布《中国企业境外可持续森林培育指南》,开创了中国海外森林采伐的新模式。

在 2007 年 APEC 第 15 次领导人非正式会议上,国家主席胡锦涛建议"建立亚太地区森林恢复和可持续管理网络"的内容写入了《悉尼宣言》。商务部于 2009 年 3 月出台《境外投资管理办法》和国别投资指南,加强了对外"走出去"企业的引导服务工作,将有力地促进中国企业积极稳妥地开展境外投资。2009 年 3 月,国家林业局会同商务部发布了《中国企业境外可持续经营利用指南》,积极指导和规范中国企业在海外的可持续林业活动,促进东道国林业的可持续发展,并维护我国政府负责任大国的国际形象。

在对外绿色信贷方面,中国银行业监督管理委员会于 2012 年年初发布了《绿色信贷指引》,推动银行业金融机构以绿色信贷为抓手,有效防范环境与社会风险,更好地服务实体经济。《绿色信贷指引》明确规定:"对拟授信的境外项目公开承诺采用相关国际惯例或国际准则,确保对拟授信项目的操作与国际良好做法在实质上保持一致。"

38 意见、建议

通过提供培训、技术支持和清洁技术来与国外企业建立伙伴关系以加强环境的改善；确保不以降低环境准入要求来吸引国外直接投资。

落实情况

近年来，中国对环境保护日益重视，中国环境保护法律法规、政策体系不断发展和完善，环境监督管理能力大大提高，中国在污染物排放和标准方面，国内外企业一视同仁。中国对所有外国来华投资企业严格执行各项环境管理制度，绝不允许以降低环境准入要求来吸引国外直接投资。近年来，在社会舆论的监督下，中国对在中国发生环境违法行为的外资企业，给予了一定处罚。例如，2007年原国家环保总局对日立建机（中国）有限公司违规超标排放废水行为，进行了依法查处。

2009年，商务部、环境保护部联合发布《关于加强外商投资节能环保统计工作的通知》，要求各地商务主管部门在办理外商投资企业设立和变更时，须要求企业提交环保部门审批的环境影响评价文件，并在外商投资企业审批管理系统中增设了环保指标，以提高利用外资质量，全面衡量外商投资节能环保水平。

39 意见、建议

通过引入清洁煤技术，提高能源利用效率和燃料转换来强化国内和国际合作以减少东北亚地区的跨界大气污染。

落实情况

为了改善大气环境质量，对于重要的大气污染物，中国政府在国民经济和社会发展规划纲要中将二氧化硫等主要污染物列为强制性减排指标，在具体的工程减排措施中应用了包括清洁煤技术在内的技术手段，提高了能源利用效率和燃料转换率；经过几年的努力，已经取得了明显的成效，"十一五"期间，二氧化硫排放总量减少14.29%。

几年来，通过双、多边合作，开展了加强能源利用和污染减排政策、法律、法规研究以及能力建设等方面的国际合作项目，如世界银行技援"促进中国循环经济发展的政策研究"、"建立中国绿色国民核算体系研究"等项目，在政策层面和战略高度提高了认识，倡导使用清洁能源的理念，引入并运用先进能源技术，有效提高能源利用率。

环境保护部于2008年与亚行合作开展二氧化硫排污交易体系研究，借助国际先进理念，设计二氧化硫排污交易体系，加强大气污染控制管理。与双边国家积极开展了一批节能减排的试点项目，包括：中意合作宁夏生物质能发电项目、中意环保合作-太原能效项目、中意环保合作-中国可持续发展能源效率门户网站项目以及中德合作可再生能源政策与节能项目等。

作为成员国，中国积极参与东北亚环境合作机制（NEASPEC）的各项活动，包括火电厂跨界大气污染控制、跨境自然保护等项目，并由环境保护部牵头实施"建立东北亚跨界区域自然保护协调机制"项目。

现阶段，中国污染减排的成效不仅明显改善了本国的环境质量，而且也对区域环境质量改善作出了贡献。

40 意见、建议

确保《蒙特利尔议定书》提出的逐步淘汰臭氧层损耗物质的中期和最终目标能够按照计划实现。

落实情况

自1991年正式加入《蒙特利尔议定书》以来，中国恪守议定书各项规定，通过组建履约管理机构、开展公约政策研究和谈判、国内履约政策法规建设、多边基金项目实施等活动，取得了履约工作的显著进展。2007年9月，在《蒙特利尔议定书》第十九次缔约方大会上，原国家环保总局获得联合国颁发的《蒙特利尔议定书》"优秀实施奖"，海关总署和北京奥组委分别获得"公共意识奖"。经过持续不断的努力，2007年7月1日，中国完成了全氯氟烃（CFCs）和哈龙这两类最主要的消耗臭氧层物质淘汰，比《蒙特利尔议定书》规定提前两年半完成履约目标。在此基础上，中国又于2010年1月1日成功地淘汰了四氯化碳和甲基氯仿的生产和使用。20多年来，中国共计淘汰了10万t消耗臭氧层物质的生产量和11万t的消费量，约占发展中国家淘汰总量的一半，圆满完成了《蒙特利尔议定书》阶段性履约任务。作为发展中国家中最大的消耗臭氧层物质生产和消费国，中国没有出现任何不履约的情形，成为第5条发展中国家履约的典范。

为完成《蒙特利尔议定书》中期和最终目标，中国采取了如下措施：建立了有效的实施机制、成熟的管理框架、较为完善的政策法规体系和强有力的履约队伍，履约能力得到加强。在国内履约管理框架建设方面，1991年国务院批准成立了由原国家环保总局牵头、18个部门组成的国家保护臭氧层领导小组，又于2000年成立了消耗臭氧层物质进出口管理办公室，切实加强履约工作的协调指导。1993年，国务院颁布了《中国逐步淘汰消耗臭氧层物质国家方案》，1999年对该国家方案进行修订，进一步完善了履约工作的政策措施。在政策法规制定方面，20年来，中国颁布实施了100多项相关政策法规，建立完善了消耗臭氧层物质生产、消费、产品质量和进出口管理制度，形成了以禁止新建、改建、扩建生产线，实施生产、消费、进出口配额许可证制度为核心的法律、法规和政策体系。2010年6月1日开始施行的《消耗臭氧层物质管理条例》，进一步规范了消耗臭氧层物质管理，强化了对违法行为的处罚力度，为可持续履约提供了强有力的法律保障。在探索履约管理模式方面，中国很早提出了生产削减、消费转换、替代品发展和政策法规建设"四同步"的履约管理框架，并通过"行业方式"付诸实施，取得了较好成效。中国是"行业方式"的首个践行者，通过组织实施25个行业计划，使行业内的所有企业纳入淘汰总体计划，避免了"点上治理、面上破坏"、"边治理、边污染"的顽疾，充分发挥了行业主管部门和行业协会的作用，增强了履约工作的灵活性和自主性，降低了淘汰成本，确保了履约任务的按时完成。2011年12月，环境保护部在上海举办了"加速淘汰含氢氯氟烃行业计划实施启动大会"，标志着中国消费行业含氢氯氟烃淘汰管理计划正式启动实施。

41

意见、建议

制订一个连续的气候变化的国家计划，列出目前采取的与气候变化相关的行动和拟采取的行动，以提高其综合效用和影响。

落实情况

中国制定了连续的气候变化的国家计划：2007年6月，国家发展和改革委员会发布了《中国应对气候变化国家方案》，明确了到2010年应对气候变化的具体目标、基本原则、重点领域及政策措施。根据该方案，中国将采取一系列法律、经济、行政及技术等手段，减缓温室气体排放，并提高适应气候变化的能力；自2008年开始，国务院新闻办公室发表《中国应对气候变化的政策与行动》白皮书，全面介绍了气候变化对中国的影响、中国减缓和适应气候变化的政策与行动、以及中国对此进行的体制机制建设，以后每年都发布该白皮书。

2009年11月，中国政府公布了控制温室气体排放的行动目标，决定到2020年单位国内生产总值二氧化碳排放比2005年下降40%～45%。《中华人民共和国国民经济和社会发展第十二个五年规划纲要》提出到2015年，单位国内生产总值能源消耗和二氧化碳排放分别降低16%、17%。

2011年12月，国务院印发了《"十二五"控制温室气体排放工作方案》。该方案提出"十二五"期间，要大幅度降低单位国内生产总值二氧化碳排放，到2015年全国单位国内生产总值二氧化碳排放比2010年下降17%。控制非能源活动二氧化碳排放和甲烷、氧化亚氮、氢氟碳化物、全氟化碳、六氟化硫等温室气体的排放。进一步完善应对气候变化政策体系、体制机制，基本建立温室气体排放统计核算体系，逐步形成碳排放交易市场。开展低碳试验试点，推广一批具有良好减排效果的低碳技术和产品。

42 意见、建议

通过控制陆基污染源，进一步努力保护和改善海岸带和邻近海域水质，完善水产业的环境管理法规和政府监督机制。

落实情况

为控制陆源污染，保护和改善包括重点海域在内的海洋环境质量，逐步建立了中国保护海洋和海岸带环境的法律法规体系。2008年修订的《中华人民共和国水污染防治法》为开展陆基污染源控制提供了基本依据，为工业、农业尤其是水产养殖等重点行业污染控制明确了管理重点。2008年中国又针对捕鱼、商业海运、娱乐活动、客运船舶等海洋产业的垃圾管理制定了行业指南。2009年9月2日国务院第79次常务会议通过《防治船舶污染海洋环境管理条例》并于2010年3月1日起实施。

为落实各项法律法规的具体要求，中国政府提出了"海陆统筹、河海兼顾"的海洋环境治理方针，提出了让江河湖海休养生息的保护战略。中国政府自2001年开始实施碧海行动计划，2006年以来，又启动了《中国保护海洋环境免受陆基活动影响国家行动计划》编制工作，将国家目标与全球和区域目标紧密结合。同时，环境保护部联合各部委持续开展全国海洋环境保护联合执法检查工作。截至2010年，联合执法检查主体已经扩展到9个部门，包括环境保护部、国家发展和改革委员会、监察部、财政部、住房和城乡建设部、交通运输部、农业部、国家海洋局、全军环办。年度执法检查工作的开展为落实海洋环境保护各项法律法规和政策措施，强化海洋环境保护提供了基本保障。

中国环境保护"十一五"规划明确提出海洋环境保护的主要目标是编制和实施以削减陆源污染物排放为重点，以重点海域污染治理为突破口，加强海洋生态保护，提高海洋环境灾害应急能力，改善海洋生态系统服务功能。国家环境保护"十二五"规划在"十一五"基础上，又提出了对海洋生物多样性进行保护的要求，同时，要求建立海洋环境监测数据共享机制，为全面深入开展海洋环境保护提供基础。2012年5月，环境保护部、国家发展和改革委员会、财政部、水利部四部委联合发布关于印发《重点流域水污染防治规划(2011—2015年)》的通知（环发[2012]58号），在陆基污染物控制方面，明确提出要根据海洋功能区划定禁排区和禁倒区，对不符合要求的排污口和倾倒区限期关闭。同时，积极推进沿海区域综合整治与修复，先后开展了一系列入海河流综合整治工程，有效实施了一批岸线、海岛、滨海湿地和珊瑚礁、红树林等重要生态系统的修复工程。

目前，中国政府正在组织编制《近岸海域污染防治"十二五"规划》，这是中国第一个全国性海域污染防治规划。同时，国家海洋局依据《海洋环境保护法》和"三定"方案的要求，严格监管陆源污染排放，优化排污口布局，对设置不合理、排放有毒污染物的陆源入海排污口实施关停并转，制定排污口入海排放标准，设置主要排污口入海监测断面，对超标排放污染物的排污口加强监管实现达标排放，探索建立重点海域污染物总量控制制度。

43 意见、建议

在中国的发展合作计划中，系统地考虑环境问题。

落实情况

环境问题已经成为中国在制定与实施发展合作计划时考虑的重要方面，商务部从"十一五"开始，逐步调整了中国的商务发展战略，一是在发展观念上，强调对外资引进质量和结构的要求，从技术含量、国内配套比例、资源消耗、环境保护、新增就业等方面建立符合科学发展观要求的吸收外资考核评价体系；二是在引资结构上，鼓励引入技术辐射能力强、吸收就业能力强、资源节约型的外资企业，鼓励外资投向现代农业、高技术产业、基础设施、新能源、节能环保产业和现代服务业等领域。

在国际履约方面，中国已经积极加入全球性30余项国际环境类公约，与50多个国家签署了双边环境保护合作协议或备忘录，在全球环境保护中发挥着日益重要的作用。为进一步提高国际履约能力，中国正在编制履行环境国际公约"十二五"专项规划、"十二五"持久性有机污染物（POPs）削减控制专项规划。

在双边和多边环境发展合作中，中国政府有计划地在环境保护能力建设、环保法规和政策、水环境管理、大气污染控制、生态保护、化学品污染防治和管理、应对全球环境问题等方面开发和开展合作项目。在合作项目开发过程中，同时考虑国内环境保护需求和合作国家的优势领域，初步建立了合作方向框架。如意大利在环境技术和监测等方面在世界上处于领先地位，合作项目主要涉及空气质量监测、节能技术和环保技术等；美国、德国在环境政策与管理、环境立法和执法方面具有优势，合作项目主要涉及环境政策咨询，以期为国内环境法律法规制定提供借鉴；澳大利亚在水质管理方面具有独到的经验，因此与澳大利亚的合作主要涉及水环境管理，包括跨行政区域流域管理、生态补偿等；瑞典在环境管理领域方面较强，因此与瑞典的合作主要涉及环境管理；挪威在应对全球环境问题方面开展的研究较多，目前与挪威的合作重点是汞污染控制和应对气候变化等领域。另外，中国与欧盟还合作实施了中欧生物多样性保护项目，有力地推进了地方层面开展生物多样性保护工作。在突发环境事件应急工作方面，中国陆续与美国、德国、俄罗斯等国家开展了合作交流。

为推动环境与发展领域的交流，中国政府自1992年就成立了"中国环境与发展国际合作委员会"，该委员会由中外环境与发展领域高层人士与专家组成，目前，该委员会由李克强副总理任主席，环境保护部周生贤部长和加拿大国际发展署比格斯署长任副主席。该委员会在交流、传播国际环境与发展领域内的成功经验、针对环境与发展领域的重大问题向中国政府领导层与各级决策者提供前瞻性、战略性、预警性政策建议，推动中国实施可持续发展战略，建设资源节约型、环境友好型社会等方面起到了积极作用。

44 意见、建议

在所有行业内将能源消耗强度的目标转换为更加富有挑战性的能源效率目标，运用综合的手段来实现这一目标，包括价格政策、需求管理、引进清洁技术、建设节能建筑、房屋和使用节能设备等。

落实情况

"十一五"以来，中国政府节能减排力度不断加强，除了能效消耗强度指标外，中国加大了对能源效率指标的考核，中国《"十一五"能源发展规划》中对重点耗能行业主要产品和耗能设备都提出了具体的控制要求。"十二五"国民经济和社会发展纲要除了能源消耗强度类指标外，还提出了能源资源效率类指标，包括非化石能源占一次能源消费比例、资源产出率、工业固体废物综合利用率等指标。为确立绿色发展的理念，提升中国工业节能发展水平，工业和信息化部发布了《工业节能"十二五"规划》，提出到2015年，规模以上工业增加值能耗比2010年下降21%左右，"十二五"期间预计实现节能量6.7亿t标准煤，并对一些主要行业、主要产品制定了节能目标。在相关部门和省级规划中，进一步强化了对能源效率类指标的考核。

为确保节能减排目标实现，中国政府采取了一系列措施和手段。综合采取包括加快节能减排工程建设、深入推动工业结构调整、淘汰电力、钢铁、焦炭、铁合金、电石、电解铝、铜铅锌冶炼、水泥、平板玻璃、造纸和煤炭等行业落后产能，加强高能耗高污染行业运营监管等手段，推动节能减排目标实现。同时，配套实施差别电价制度和惩罚性电价政策，组织对高耗能、高污染行业节能减排工作专项检查，开展对能耗限额标准执行情况的监督检查，清理和纠正各地在电价、地价、税费等方面对高耗能、高污染行业的优惠政策。实施重污染行业信息公开、强化新建建筑执行能耗限额标准全过程监督管理，实施建筑能效专项测评等，从2008年起，所有新建商品房销售时在买卖合同等文件中要载明耗能量、节能措施等信息。建立并完善大型公共建筑节能运行监管体系。

加快节能减排技术研发、示范和推广。在国家重点基础研究发展计划、国家科技支撑计划和国家高技术研究发展计划等科技专项计划中,安排一批节能减排重大技术项目、加快节能减排技术产业化示范和推广。在钢铁、有色、煤炭、电力、石油石化、化工、建材、纺织、造纸、建筑等重点行业,推广潜力大、应用面广的重大节能减排技术。

国务院发布的《"十二五"节能减排综合性工作方案》,从目标分解和考核、产业结构调整和优化、发展循环经济、完善经济激励政策、加快节能减排技术研发、示范和应用、推广包括需求管理在内的市场化机制和手段、加强节能减排基础能力建设等方面提出了综合性政策手段和工具。目前,各类手段在各个层面正得到完善和加强。

45 意见、建议

鼓励使用清洁燃料(包括洁净煤技术,洗煤和烟气脱硫)和清洁的机动车燃料。

落实情况

2003年发布的《国家清洁能源行动实施方案》提出了促进清洁能源技术应用水平的一系列措施。《中华人民共和国国民经济和社会发展第十一个五年规划纲要》提出的"十一五"期间单位国内生产总值能耗降低20%左右,主要污染物排放总量减少10%的约束性指标,促进了清洁能源的使用;同时明确提出要将非化石能源消费占一次能源消费的比例由8.3%提高到11.4%。在烟气脱硫方面,中国"十二五"污染物总量减排规划中明确规定,所有燃煤电厂必须要上烟气脱硫装置,对所有燃煤设施的脱硫效率提出强制性要求。在具体的工程减排措施中应用了包括清洁煤技术在内的技术手段,提高了能源利用效率和燃料转换率。

"十一五"期间,中国政府通过实施乘用车燃料消耗量限值标准和鼓励购买小排量汽车的财税政策等措施,先进内燃机、高效变速器、轻量化材料、整车优化设计以及普通混合动力等节能技术和产品得到大力推广,汽车平均燃料消耗量明显降低;例如,交通运输部针对营运车辆,颁布了《道路运输车辆燃料消耗量检测和监督管理办法》,对不符合限制标准的车辆不允许进入营运市场。同时大力推广天然气车辆在道路运输中的应用,在宁夏、辽宁、山东等省区开展了试点工作,2011年及2012年补助项目共替代燃料308 729.46t标准油。

"十二五"期间针对机动车污染排放,中国政府制定了《"十二五"战略性新兴产业发展规划》和《节能与新能源汽车产业发展规划》,均强调要加快培育和发展新能源汽车产业,重点推进纯电动汽车、插电式混合动力汽车产业化。以快速降低汽车燃料消耗量为目标,大力推广普及节能汽车,提升汽车产业整体技术水平。同时,为了加强机动车污染防治工作,中国制定了重型车第三、第四、第五阶段排放标准和轻型车第三、第四阶段排放标准,目前正在制定轻型车第五阶段排放标准。为了提高车用燃油清洁化水平,中国制定了第四、第五阶段车用汽油、柴油中有害物质控制标准。

46 意见、建议

制定和实施考虑环境外部性的国家交通战略，采用综合手段管理私人和公共交通；形成现代化的、发展可持续交通系统的制度框架；综合运用法规和经济手段（如税收）来激励人们选择合理的交通方式。

47 意见、建议

发展城市地区的大容量交通，采取措施鼓励城市使用清洁的交通方式（如自行车）。

落实情况

为了推动可持续交通，中国政府采取了多种政策措施，一是确立并实施公共交通优先发展战略。二是有序推进轻轨、地铁、有轨电车等城市轨道交通网络建设。三是积极发展地面快速公交系统，提高线网密度和站点覆盖率。四是规范发展城市出租车业，合理引导私人机动车出行，倡导非机动方式出行。

"十二五"时期，中国政府加强了对交通运输业节能管理和污染控制。中国政府公布的《"十二五"交通运输业发展规划》提出了营运车辆单位运输周转量能耗和二氧化碳排放下降率、营运船舶单位运输周转量能耗和二氧化碳排放下降率、民航运输吨公里能耗和二氧化碳排放下降率、国省道单位行驶量用地面积下降率、沿海港口单位长度码头岸线通过能力提高率、总悬浮颗粒物（TSP）和化学需氧量等主要污染物排放强度下降率等控制性目标，对交通运输业节能减排目标提出了具体要求。中国政府公布的《铁路"十二五"节能规划》，明确提出"十二五"时期铁路行业节能主要目标是：基本完善行业节能减排法规、政策和标准；建立完善行业节能监测体系；单位运输工作量综合能耗降低 5%，从 2010 年的 5.01t 标准煤/百万换算吨公里下降到 2015 年的 4.76t 标准煤/百万换算吨公里。

中国政府高度重视城市交通发展和节能减排，"十一五"以来，为贯彻实施"绿色交通"战略，普遍实施了发展公共交通和大容量交通、鼓励清洁能源汽车发展战略。出台老旧汽车"以旧换新"经济补贴政策，加速了高污染车辆的更新淘汰。同时，为了降低城市交通拥挤，降低城市交通污染，北京、上海、广州等地提前实施了更严格的机动车排放标准和车用燃料标准，出台了机动车牌拍卖/摇号政策，北京和上海等城市实施差别停车费，提高市中心等交通拥挤地区停车收费标准。推进城市公共自行车出行系统建设，交通运输节能减排专项资金对杭州等地的公共自行车进行了补助。

在国务院发布的《"十二五"节能减排综合性工作方案》中，明确提出推进交通运输节能减排。加快构建综合交通运输体系，优化交通运输结构。积极发展城市公共交通，科学合理配置城市各种交通资源，有序推进城市轨道交通建设。提高铁路电气化比重。实施低碳交通运输体系建设城市试点，深入开展"车船路港"千家企业低碳交通运输专项行动，推广公路甩挂运输，全面推行不停车收费系统，实施内河船型标准化，推进航空、远洋运输业节能减排。开展机场、码头、车站节能改造。加速淘汰老旧汽车、机车、船舶，基本淘汰2005年以前注册运营的"黄标车"，加快提升车用燃油品质。实施第四阶段机动车排放标准，在有条件的重点城市和地区逐步实施第五阶段排放标准。全面推行机动车环保标志管理，探索城市调控机动车保有总量，积极推广节能与新能源汽车。

48

意见、建议

继续努力为农村人口提供安全的供水和卫生设施以实现国内目标和国际承诺（例如：千年声明和可持续发展世界首脑会议）；继续安装水表和征收水费，并将社会因素考虑进去。

落实情况

2005年安排投资解决了1 104万农村居民的饮水安全问题。"十一五"期间，中国农村供水和卫生设施得到较大发展。2006—2010年国家已安排总投资1 053亿元，其中中央投资590亿元，地方自筹配套439亿元，用于解决2.1亿人的饮水安全问题。2005—2010年，中国累计解决了2.21亿农村人口的饮水安全问题。"十一五"期间，国家通过中央财政累计投入39.4亿元，补助农村地区修建无害化卫生厕所1 083.2万座。截至2010年年底，中国农村卫生厕所普及率已达67.3%。通过开展农村改厕工作，有效预防和减少了肠道传染病和寄生虫病的发生，明显改善了农村环境卫生面貌，促进了群众卫生行为和健康习惯的养成，得到了人民群众的认可。

"十二五"时期，国家发展和改革委员会、水利部、住房和城乡建设部联合发布《水利发展规划（2011—2015年）》，明确提出到2015年，全面解决2.98亿农村人口和11.4万所农村学校的饮水安全问题。《国务院关于印发国家基本公共服务体系"十二五"规划的通知》提出在"十二五"期间要加强饮用水卫生监督监测体系建设。自2011年起，国家将卫生监督协管服务纳入基本公共卫生服务项目，对农村集中式供水、学校供水等开展卫生安全巡检工作，掌握农村饮用水卫生管理状况，扩大卫生监督服务覆盖范围，进一步提高农村饮用水卫生安全保障能力和水平。

《国家环境保护"十二五"规划》强调，要开展农村饮用水水源地调查评估，推进农村饮用水水源保护区或保护范围的划定工作。强化饮用水水源环境综合整治。建立和完善农村饮用水水源地环境监管体系，加大执法检查力度。开展环境保护宣传教育，提高农村居民水源保护意识。在有条件的地区推行城乡供水一体化。

征收水费作为加强水资源管理和提高水污染治理效率的重要手段，正在全中国范围内得到扩展应用。截至 2010 年年底，全国大多数城市住户已经安装了水表，目前，作为基本服务均等化的一个重要建设内容，水表安装正逐步向农村广大区域延伸。在水资源费和污水处理收费政策方面，由于中国经济发展水平和发展阶段的区域差异，东部发达地区正逐步建立面向市场的阶梯水价制度，而在中西部地区以及欠发达区域，中国政府则实施水价补贴政策。

49

意见、建议

通过实施林业管理计划，颁发植树证明和林产品标志来促进可持续的林业管理；在林业方面，扩大与供给国的合作，保证进口的木材和木材产品来自管理良好的、可持续的林区。

落实情况

中国正在采取积极措施以实现林业的可持续管理。2007 年 8 月，国家林业局、国家发展改革委、财政部、商务部、国家税务总局、中国银行业监督管理委员会、中国证券监督管理委员会联合发布了《林业产业政策要点》。其政策目标是"全面落实科学发展观，实施以生态建设为主的林业发展战略，发挥市场配置资源的基础性作用和国家的宏观调控作用，逐步建立起门类齐全，优质高效，竞争有序，充满活力的现代林业产业体系，充分发挥林业的多种功能，大力提升林产品的供给能力，最大限度地满足经济社会发展对林产品与服务的多样化需求"。2009 年以来，中国加强了森林抚育工作。国家林业局先后修订发布了《森林抚育规程》。在《林业发展"十二五"规划》、《全国造林绿化纲要（2011—2020 年）》中明确了森林抚育规划任务、强化了对森林抚育经营工作的指导。从 2009 年开始，中央财政共安排森林抚育资金 133.06 亿元、抚育任务 12 254.88 万亩，对提高森林质量、提升森林生态系统综合服务功能等都产生了积极作用。

2007 年 8 月，国家林业局会同商务部联合编制发布了《中国企业境外可持续森林培育指南》。要求中国企业在境外从事森林培育活动，要自觉遵守国际公约尤其是中国政府已加入的相关国际公约，执行中国针对企业境外投资活动的方针政策，严格执行所在国有关法律法规，依法保护林地，严格保护高价值森林，严禁非法转变林地用途等。

2008 年 5 月，中美两国政府签署了《关于打击木材非法采伐及相关贸易谅解备忘录》，建立了中国国家林业局、外交部、商务部和海关总署，美国贸易代表办公室、国务院、司法部、鱼和野生动物局、海关与边境署、林务局等组成的中美打击非法采伐及相关贸易双边论坛。

2009年5月，国家林业局、商务部又联合发布《中国企业境外森林可持续经营利用指南》，从森林资源经营、木材加工与运输、人员培训与技术指导、建立多利益方的公示和咨询制度、加强环境保护与生物多样性保护、促进当地社区发展等方面提出了规范性要求。

50 意见、建议

评估能源、水和其他自然资源的价格水平，以更好地反映它们的稀缺价值，内部化其外部性；建立一种机制来补偿或减少由于价格提高所可能带来的对贫困地区人口的负面影响。

落实情况

中国政府非常重视自然资源定价政策的制定和实施，以更好地反映其稀缺价值，并减少其环境外部性。中国政府近年来一直在致力于推动资源税改革，自2010年6月资源税改革试点在新疆启动以来，截至2010年年底，资源税改革试点扩大到西部12个省份，原油、天然气资源税由原来的从量计征改为从价计征。2011年9月，国务院出台《关于修改〈中华人民共和国资源税暂行条例〉的决定》，明确资源税的应纳税额计征方式；10月，国务院出台《资源税暂行条例实施细则》，这意味以中国油气资源税从量计征改为从价计征为核心内容的资源税改革开始从试点地向全国全面推开，进一步完善了资源性产品价格形成机制。

针对资源税存在的税目范围过窄、税负较低等问题，中国政府制定的《"十二五"节能减排综合性工作方案》等一些政策文件也明确提出"十二五"要继续推进资源税费改革，扩大资源税改革实施范围等要求。总之，中国将加大资源税改革步伐，提高税率，扩大征收范围，把更多的资源性产品列入其中。

中国政府高度重视水资源定价政策改革，2009年7月6日，国家发展和改革委员会与住房和城乡建设部联合发布《关于做好城市供水价格管理工作的通知》，加强城市供水价格管理；2010年12月，国家发展改革委出台《城市供水定价成本监审办法(试行)》，加强城市供水定价成本审核；同年出台了《关于做好城市供水价格调整成本公开试点工作的指导意见》，加快推进城市供水水价改革试点。中国的大部分地区，如上海、重庆、江苏、山东、宁夏等都上调水价，以解决供水成本与水价倒挂、水价未能反映水资源的稀缺程度和水环境治理成本的问题。

中国政府高度重视水价上调对低收入和贫困地区人口的影响。在《关于做好城市供水价格管理工作的通知》中，中国政府要求在水价改革过程中，要充分听取社会各方面的意见，提高水价决策的透明度。要严格履行水价调整程序，依法履行水价调整听证制度，要求各地区在调整水价时要充分考虑低收入家庭的承受能力，要切实做好对低收入家庭的保障工作，减少水价调整对低收入家庭的影响，结合价格总水平变化，对低收入家庭因地制宜地采取提高低保标准、增加补贴等多种方式，保障其基本生活水平不降低。

51 意见、建议

继续将国家和区域荒漠化防治的努力置于优先地位。

落实情况

中国正在努力防治国家和区域荒漠化。进入21世纪以来，中国采取了一系列行之有效的措施推进荒漠化防治工作。2002年颁布实施了《中华人民共和国防沙治沙法》，2005年国务院批复了《全国防沙治沙规划（2005—2010年）》，先后实施了一批防沙治沙重点工程。2007年3月，全国治沙大会召开，国务院总理温家宝、副总理回良玉出席并做了重要讲话；会上，国家林业局与防治任务较重的12个省、自治区和新疆生产建设兵团签订"十一五"防沙治沙目标责任书，进一步强化防沙治沙重点省、自治区各级人民政府责任制，明确"十一五"防沙治沙工作任务目标及应采取的措施和承担的责任。

中国是《联合国防治荒漠化公约》缔约国，积极参加相关国际组织和公约框架下开展的荒漠化防治评估和履约绩效评估工作。在2007年召开的荒漠化公约履约审查委员会会议上，中国专门做了荒漠化监测经验介绍，联合国粮农组织因此将中国选为全球土地退化评估项目的六个示范国之一。2010年以来，参加荒漠化公约影响指标试点，采用荒漠化公约规定绩效指标和影响指标体系评估履约绩效，为亚洲国家作出示范。

2007年8月，中国省级荒漠化和沙化土地监测信息管理系统建成并投入使用。2008年5月1日起中国防沙治沙行业的第一个国家标准《防沙治沙技术规范》开始实施。2009年3月，第四次全国性荒漠化和沙化监测启动，监测范围涉及30个省（自治区、直辖市）近900个县，面积470多万 km^2。2011年1月，中国政府公布了本次监测结果，截至2009年年底，全国荒漠化土地面积262.37万 km^2，沙化土地面积173.11万 km^2，分别占国土总面积的27.33%和18.03%。相关部门制订计划，加大实施力度，5年间全国荒漠化土地面积年均减少2 491km^2，沙化土地面积年均减少1 717 km^2。监测表明，中国土地荒漠化和沙化整体得到初步遏制。2009年国务院批准了由国家林业局会同有关部门制定了《省级政府防沙治沙目标责任考核办法》。2011年，国家林业局会同有关部门组织开展了"十一五"省级政府防沙治沙目标考核。目前，国家林业局正会同有关部门编制新阶段《全国防沙治沙规划》。

国家林业局编制印发《全国防沙治沙综合示范区建设规划（2011—2020年）》，明确提出了"十二五"及今后一段时期示范区建设的指导思想，将坚持科学防治、综合防治、依法防治方针，通过全面推进和加快示范区建设，力争用10年时间，实现示范区可治理沙化土地治理率达到70%，在生态改善、技术创新、政策机制、产业发展等方面建成一批防沙治沙示范样板项目。

Preface

Since the reform and opening-up of the country, China has made remarkable progress in economic and social development, achieving an average annual GDP growth rate of 9.8% between 1978 and 2010, while its economic aggregate is currently ranked second in the world. Both comprehensive national power and people's living conditions have improved significantly. However, some issues still exist in the economic growth mode of China, such as "high input, high consumption and high emissions, lack of co-ordination, difficulty in carrying out recycling, and low efficiency". Resources and the environment have paid the costs for such rapid economic development. These serious resource and environment issues will accordingly restrict economic and social development, be harmful to people's health, and endanger public safety and social harmony. If environmental issues are not handled properly, China's long-term interests will be fundamentally harmed. Therefore, transforming the mode of economic development and realising green growth is not only the strategic choice and inevitable need for China's long-term development, but also will positively contribute to global sustainable development, which will have a significant influence on human development.

As one of the advocates of green development, the OECD has put forward the necessary conceptual framework and monitoring and evaluating tools for its implementation; especially since 1992, environmental performance evaluation has played a positive role in helping participant countries to understand their environmental problems and implement measures to resolve them, such as enhancing the utilisation efficiency of energy and resources, and the implementation of environmental management policies.

In October 2005, the former SEPA (State Environmental Protection Administration) of China and the OECD have signed an agreement on co-operating in the environmental performance review of China; the conclusions and recommendations of the OECD Report on the Environmental Performance Review in China was issued in Beijing in November 2006, and the complete Report was released in Beijing on 17 July 2007. The Report conducted a comprehensive review and evaluation of the environmental performance of China in the last 10 years, focusing on the utilisation of energy and resources, ecological protection and management, and environmental and economic integration. It gave praise for the achievements China has made, saying that the Chinese government has recognised the problems of environmental degradation thus, established a comprehensive policy system, and achieved remarkable results in many fields of environmental protection. Meanwhile, the Report also pointed out some key issues that need to be resolved in the future development of China. It considers that the main problems in the current system of environmental protection in China lie in the effectiveness and efficiency of its environmental policy, which shall be urgently improved. It also suggested improving the capacities of the departments of environmental protection at various levels for inspection, supervision and enforcement, expanding the application scopes of various economic means, and strengthening the environmental liability of local governments, as well as bringing environmental factors into comprehensive economic decisions. Finally, the Report makes 51 Recommendations in the field of environmental management to assist the Chinese Government to enhance environmental performance within a sustainable development perspective.

The Chinese Government is intensely aware of the positive meaning of performance management for improving government efficiency and strengthening the effects of policy implementation. In the environmental field in particular, the Chinese Government has set COD and SO_2 as two binding indexes and listed them in the Outline of the Plan for National Economic and Social Development for the first time, which implied an important variation in the Chinese Government's awareness of environmental protection since the 11th Five-Year Plan. As for the implementation results, it is shown that the 11th Five-Year Plan's objectives and key tasks of environmental protection were fully completed by the end of 2010; the emissions of COD and SO_2 had decreased by 12.45% and 14.29% respectively by 2005, meaning that the target objectives were met well within schedule. In addition, pollution treatment facilities have been rapidly increased with the treatment rate of urban sewage raised to 72% from 52% in 2005; the percentage of installation capacity of thermal power with desulphurisation has increased from 12% to 82.6%. Meanwhile, environmental quality has improved to a certain degree: at the state-controlled monitoring section, the proportion of surface water of a quality better than Grade III has increased to 51.9% in China as a whole, while the national average urban SO_2 concentration fell by 26.3%. These achievements mainly resulted from the implementation of environmental planning and policies, strengthened foundation capacities of environmental supervision, and the comprehensive use of a variety of policy means during the 11th Five-Year Plan period. Environmental protection has become an important focus to transform the economic growth mode and promote green development and ecological progress, which is playing an increasingly significant role in the comprehensive, co-ordinated and sustainable development of the economy and society.

In order to further promote the feedback of 51 comments and recommendations from OECD, Department of Total Pollutants Control, Ministry of Environmental leading and cooperating with Chinese Academy for Environmental Planning has spent three years to compile "Feedback on 51 Comments and Recommendation of OECD China Environmental Performance Review" (hereinafter referred to as "Feedback") which has twice call for opinions from other departments of the Ministry of Environmental Protection, National Development and Reform Commission, Ministry of Housing and Infrastructure, State Forestry Administration, Ministry of Commerce and other ministries. To fully reflect China's environmental performance during the "11th Five-Year Plan" Department of Total Pollutants Control, Ministry of Environmental leading and cooperating with Chinese Academy for Environmental Planning has drafted "China Environmental Protection Status and Development –Interim Review of OECD Environmental Performance Review of China Report". The report has drawn pressure - state - response (PSR) framework proposed by OECD to present environmental pressures and development goals as well as Chinese government's efforts and prospects , and conducted a comprehensive review on the environmental performance of Chinese government during the 11th Five-Year Plan".

In 10th of October, 2012 Organization for Economic Cooperation and Development (OECD) has invited the Chinese delegation led by Wang Jinnan vice president of Chinese Academy for Environmental Planning went to Paris, France to participate in an OECD Environmental Performance group meeting. The Chinese delegation has presented the important environmental policy since 2007 OECD Environmental Performance Review in particularly during "11th Five-Year Plan" as well as feedback on 51 policy recommendations made by OECD and they are implementation. This meeting was a showcase on China's environmental performance during "11th Five-Year Plan". OECD has made highly appreciation and recognition on the work has been done.

To further promote the Chinese government's efforts on environmental performance during "11th Five-Year Plan" we drafted "Feedback on 51 Comments and Recommendation of OECD China Environmental Performance Review" and it has received the feedback from other departments of the Ministry of Environmental Protection, National Development and Reform Commission, Ministry of Housing and Infrastructure, State Forestry Administration, Ministry of Commerce and other ministries; they are provide informative material to support our works. Each MEP department, in particularly Deputy Director Yu Fei of Department of Total Pollutants Mao Yuru, Statistics Division Chief have reviewed feedback; Tu Ruihe, Deputy Director, International Cooperation Department and director Liu Ning and director Gu Li have provide strong support for the delegation head to OECD Group Meeting. Qian Yong, Director of the General Office has proposed suggestions from completeness and accuracy perspective. Also, special thanks to President Hong Yaxiong, Chinese Academy for Environmental Planning for his support on environmental performance assessment works. And Wang Jinnan who led a delegation to participate in the OECD Environmental Performance Review of Interim Group Meeting and proposed related guidance material for the publication of the book is sincerely appreciated.

The Authors

July 2013

1. Environmental stress and development objectives

1.1 Environmental stress as a result of economic and social development

Since 2000, China's economy has stepped into the trajectory of rapid growth, and China's GDP rose from 10.97 trillion CNY in 2001 to 40.12 trillion CNY in 2010, with an average annual economic growth rate of 10.5%. In 2008, China's GDP overtook Germany's as the third economy in the world. In 2010, China's GDP reached to 5.88 trillion CNY, overtaking Japan as the second-largest economy in the world. While major world economies are facing negative growth or stagnation, China's economy still maintained a relatively high growth rate, which has made great contributions to global economic recovery.

The quality of population, demographic pattern and population distribution significantly affect socio-economic stability and sustainable development. Compare with the total amount of population China is still the largest country in the world, the natural population growth rate maintained at 5‰. In 2010 China's total population reached to 1.34 billion which is 19% of overall world's population. The average life expectancy has increased from 68 at beginning of 1980s to 73.5 presently. The continuously increased population put huge pressure on sustainable development.

First of all, population growth intensifies the dichotomy between population and cultivated land. China's forest coverage rate is still lower than the global average; rural population growth and the shortage of energy caused increasing demands for food and land. Population growth intensified the imbalance between supply and demand of water resources in China. Compared with the 1990s, urban air pollution has been improved to a great extent.

However increasing coal consumption is the major driving factor of China's air quality deterioration. The water environment situation has improved to a certain degree, but is still grim. Eutrophication of lakes and reservoirs in China is increasingly aggravated. Recently the wetland's resources of China are facing threats and biodiversity is decreasing.

1.2 China's objectives of sustainable development

1.2.1 Guiding ideology and overall objectives of China's sustainable development strategy

The overall objective for the sustainable development strategy upheld by the Chinese government is to effectively control the population, significantly improve quality of life, significantly enhance levels of science and technology and education, continuously improve people's living conditions, more reasonably develop and utilise energy and resources, significantly improve ecological and environmental quality, and constantly enhance sustainable development capacity, so as to establish the co-ordinated development of the economy, society and population, as well as resources and the environment.

1.2.2 General thinking of China's sustainable development strategy

The overall strategy including five major components: Making economic restructuring as the major measure for sustainable development strategy; making the guarantee and improvement of people's livelihood as a main purpose of sustainable development strategy; making accelerating poverty alleviation as an urgent task for sustainable development strategy; making construction of a resource-saving and environment-friendly society as a focus for sustainable development strategy; making the comprehensive improvement of sustainable development capacity as a basic guarantee for sustainable development strategy.

In the "12th Five-Year Plan" period, Chinese government strengthen both social and economic development and environmental and resources objectives. It proposed 12 objectives, including: maintaining 12 million km^2 arable lands; decrease water consumption intensity (per industrial-added value) by 30%; increase agricultural efficiency rate to 0.53; rise non-fossil fuel consumption in the primary energy supply by 11.4%; reduce overall energy consumption per GDP by 16%; reduce CO_2 releasing per GDP by 17%; major pollutants reduction goals are: COD, SO_2, NH_3, NO_x at 8%, 8%, 10% and 10% respectively; forestry coverage rate will be increased to 21.66% and forestry volume will be increased to 600 million m^2.

2. Efforts of the Chinese government

2.1 Laws and regulations for environmental protection

At present, a total of 50 laws and regulations on energy/resource management and environmental protection have been issued in China, including: Constitution of the People's Republic of China, Environmental Protection Law of the People's Republic of China, Electricity Law of the People's Republic of China, Coal Industry Law of the People's Republic of China, Energy Conservation Law of the People's Republic of China, Renewable Energy Law of the People's Republic of China, Mineral Resources Law of the People's Republic of China, Water Pollution Prevention and Control Law of the People's Republic of China, Atmospheric Pollution Prevention and Control Law of the People's Republic of China, Cleaner Production Promotion Law of the People's Republic of China, Circular Economy Promotion Law of the People's Republic of China, and Law of the People's Republic of China on Prevention of Environmental Pollution Caused by Solid Waste, and so on.

Since the 11th Five-Year Plan, the Chinese Government has constantly improved its environmental laws and regulations in the industrial pollution and energy fields; it revised and issued Renewable Energy Law of the People's Republic of China (amendment); formulated and implemented Circular Economy Promotion Law of the People's Republic of China; and successively promulgated seven administrative regulations, such as Regulation on Planning Environmental Impact Assessment, and Administrative Regulations on Recycling and Disposing Waste Electrical Appliances and Electronics.

2.2 Environmental standards and planning

The current standard system consists of five categories of standards divided into the national and local levels, namely standards for environmental quality, standards for pollutant discharge, specifications for environmental monitoring (standards for environmental monitoring analysis method, standards for environmental samples, and technical specifications for environmental monitoring), management standards and primary standards of environment (basic standards for environment and technical specifications for formulation and revision of standards). The levels of about 100 standards for environmental protection are raised annually on pollution reduction and improving environmental quality. At the end of the 11th Five-Year Plan, China had issued a total of 1 494 standards for environmental protection; 1 367 enforcing atandards, 14 national standards for environmental quality, 138 national standards for pollutants discharge, 705 specifications for environmental monitoring, 437 management standards, and 18 basic standards of environment.

It is of special note that the Ministry of Environmental Protection conducted the third revision and issued a new Ambient Air Quality Standard (GB 3095—2012) on 29 February 2012 in order to control and improve air quality. The new version adjusted the classification for environment air function areas and cancelled the third function area; modified items of pollutant and monitoring specification; introduced limits for $PM_{2.5}$ and O_3; set stricter concentration limits of NO_2 and PM_{10}, etc. with annual average Class II concentration of NO_2 and PM_{10} lowered from 0.05 mg/m^3 and 0.1 mg/m^3 to 0.04 mg/m^3 and 0.07 mg/m^3 respectively; revised the provisions for validity of data statistics; and increased the appendix for limits of other designated pollutants.

Table 2.1 Concentration limits of pollutants in ambient air   unit: mg/m^3

No.	Item	Average time	Concentration limit	
			Class I	Class II
1	SO_2	Annual mean	0.020	0.060
		Average for 24 hours	0.050	0.150
		Average for 1 hour	0.150	0.500
2	NO_2	Annual mean	0.040	0.040
		Average for 24 hours	0.080	0.080
		Average for 1 hour	0.200	0.200
3	CO	Average for 24 hours	4.000	4.000
		Average for 1 hour	10.000	10.00
4	O_3	Average for 8 hours with top concentration daily	0.100	0.160
		Average for 1 hour	0.160	0.200
5	PM_{10}	Annual mean	0.040	0.070
		Average for 24 hours	0.050	0.150
6	$PM_{2.5}$	Annual mean	0.015	0.035
		Average for 24 hours	0.035	0.075

In addition, the Chinese Government has newly revised the Standard for Drinking Water Quality (GB 5749—2006), which stipulates the sanitary requirements for potable water quality, quality of potable water sources, the centralised water supply unit, secondary water supply and security products involving potable water, as well as water quality monitoring and inspection methods, among which, the indexes of water quality monitoring increased from the original 35 items to 106 items (an increase of 71 items), and revised 8 items. This Standard has been effective since July 1st 2007, providing a basic guarantee and the basis to improve the safety of water supply.

In order to implement the relevant laws, regulations and standards for energy, resources and the environment, the Chinese Government established a planning hierarchy; the Outline of Plan for National Economic and Social Development is the most important planning at the national level, which involves the various fields of China's national economy, including economic development, population development, social security, industrial policies, resources and energy utilisation; it conducts the overall planning and deployment, and puts forward the specific objectives and corresponding controlling indexes for various fields, with five-years as one cycle.

In order to form a co-ordinated development pattern among population, economy, resources and environment, the Chinese Government promulgated the National Planning for Major Function Zones in 2010, which proposes that China's land and space is divided into optimisation development zones, key development zones, limited development zones and prohibited development zones, according to the development plan: optimisation development zones are mainly to speed up the transformation of the economic development mode, adjust and optimise economic structure, and participate in the global division of labour to promote competition levels; key development zones are mainly to promote sustainable economic development, promote its new industrialisation process and enhance the industrial cluster capacity on the basis of optimising structure, improving efficiency, reducing the consumption of resources and protecting the environment; limited development zones mainly refer to the major producing areas of agricultural products, focusing on protecting cultivated land, cultivating modern agriculture, enhancing the comprehensive agricultural production capacity, and increasing farmers' incomes; prohibited development zones refer to the key ecological function areas, which need to restrict large-scale and intensive industrialisation and urbanisation during the national and spatial development, so as to maintain and improve the capacity of ecological products provision. Planning also required adjusting and improving the related programme, policy, laws and regulations on the finance, investment, industry, land, agriculture, population, and environment, etc., and establishing and improving the performance evaluation system.

In the environmental field, China has basically formed a set of environmental planning systems with Chinese circumstances after 30 years of development. The planning system is mainly divided into the national and local levels. National planning consists of four categories: first is the national five-year plan for environmental protection, which is an overall national plan to define the objectives and indexes, and the main tasks and measures for environmental protection at the national level. Second is the national special planning for environmental protection, which is mainly to solve the prominent problems in the field of environmental protection. Third is the national special planning related to environmental protection involving the MEP, which reflects the link between environmental protection and resource exploitation and planning for social and economic development. And fourth is development planning from the departments of environmental protection, which aims mainly to strengthen the responsibilities and capacity building of the departments of environmental protection. Local planning includes three categories: regional planning, provincial planning and prefecture-level planning for environmental protection.

Since the 11th Five-Year Plan, with the gradual deepening of the state and the public's understanding of environmental and economic issues, significant changes have occurred in environmental protection, from awareness to practice; environmental planning is gradually eliminating the dilemma that used to be subordinate planning of social and economic development. The 11th Five-Year National Planning for Economic and Social Development firstly listed the environmental indexes into the planning index system as binding indexes, which highlighted the importance of environmental problems in the new period. The role of environmental planning is greatly enhanced during implementation of the state strategy for environmental protection, promoting and carrying out "three historic transformations" of environmental protection, and strengthening environmental protection combined with the macroeconomic control, as well as optimising economic growth.

2.3 Environmental management system and mechanisms

In order to strengthen national environmental protection, the Ministry of Environmental Protection (MEP) was formally established in March 2008, with successively additions of Department of Total Pollutants Control, Department of Environmental Monitoring, Department of Education and Communication, Department of Nuclear Safety Management (Department of Radioactive Safety Management), and Centre of Satellite Environment Application, etc. The newly established MEP strengthens the overall planning and co-ordination for major issues, such as total emission reduction, environmental supervision, nuclear and radiation supervision. With the establishment of the MEP, each province has upgraded the provincial Environmental Protection Bureau to the provincial Department of Environmental Protection.

In order to strengthen the national environmental regulatory, the former SEPA formed six regional supervision centres for environmental protection – North China, East China, South China, North-West, South-West, and North-East China – as well as six regional supervision stations for nuclear and radiation safety: the Northern, Shanghai, Guangdong, Sichuan, North-East and North-West, since 1999; among them, the former are agencies of the MEP, whose main functions are to supervise local implementation of national environmental policies, planning, regulations and standards; to investigate cases of serious environment pollution and ecological destruction, and co-ordinate and deal with significant trans-provincial, basin and coastal environmental issues. Some provinces and cities have also established independent regional supervision centres; for example: Jiangsu Province has three separate regional supervision centres for environmental protection for south Jiangsu, middle Jiangsu and north Jiangsu; Shaanxi Province has also established the North Shaanxi Supervision Centre of Environmental Protection.

Through organisational restructuring, the environmental protection management system has been streamlined and a basic environmental administration system of "national supervision, local regulatory and responsible entity" set up.

2.4 Input and capacity building for environmental protection

The past 10 years have been the period of China's largest growth in input into environmental protection;

During the 11th Five-Year Plan period, all governments from the central to the local level have increased their financial support for environmental protection. Central Finance supported environmental protection produced the best record in history: the total allocated budget capital amounted to 10.03 billion CNY, 4.71 times that in the 10th Five-Year Plan Period. Overall central investment into environmental protection was 156.4 billion CNY, nearly three times that of the 10th Five-Year Plan period. Accordingly, 2.1 trillion CNY of the whole social budget was input into environmental protection, which effectively promoted the construction of environmental protection infrastructures.

Through strengthening four aspects of capacity building, i.e., the automatic monitoring of national key enterprises under supervision, supervisory monitoring of pollution sources, environment supervision and enforcement, and environmental information and statistics, MEP is attempting to establish a set of scientific systems for the transmission, verification and analysis of total amount data of major pollutants; a set of environment monitoring systems combined with the supervisory monitoring of pollution sources and on-line automatic monitoring of key pollution sources; and a set of strict and feasible evaluation systems for total emission reduction.

2.5 Environmental policies and measures

2.5.1 Performance management

In order to reach each goal of environmental protection policy, the Chinese Government has established a series of performance management instruments in the environmental protection field; implemented a target responsibility system and a performance appraisal system in aspects of pollution reduction and the comprehensive improvement of the urban environment, key basin water quality target assessment and eco-environmental quality assessment, etc., and strengthened the environmental performance management of regional and local governments. In 2011, the Chinese Government launched the pilots for pollution reduction and performance management, further deepening the application of performance management tools in pollution reduction.

Actively promoting the comprehensive improvement of the urban environment and quantitative assessment, the national 661 cities have been included in the city test range with more than 20 million basic data on cities' environmental management, implemented the on-line transmission and system check, adopted various instruments, such as the mutual review and spot check at the provincial level, and carried out telephone surveys on public satisfaction according to international practices.

In creating national environmentally-friendly model cities as an incentive policy for environmental protection with Chinese characteristics, the aim is to fully mobilise the enthusiasm of local governments to solve the major environmental problems affecting the public, increase input into urban environmental protection, strengthen the environmental infrastructure, accelerate the optimisation of the industrial structure, and speed up the improvement of urban environmental quality. In 2010, the percentage of annual days with excellent air quality and the compliance rate of water quality of surface water environmental function areas of the national model cities were respectively 21.91% and 11.36%, higher than the national averages; the rates of centralised disposal of medical waste, sewage, and the hazard-free treatment of household waste were respectively 14.91%, 24.74% and 24.27%, higher than the national averages. Investigation by the National Bureau of Statistics indicates that public satisfaction of the environment in the national environmentally-friendly model cities is far higher than that in other cities.

2.5.2 Emission Control of gross pollutants

Total-amount control for pollutant discharge is one of the most important environmental policies in China, which includes a series of measures and means. From the national level, the Outline of the 11th Five-Year Plan on National Economic and Social Development stipulated the requirements for quantisation control of main pollutant emission. In order to support the achievement of such objectives, the State Council specially formulated the Work Scheme of the Comprehensive Energy Conservation and Emission Reduction to conduct overall deployment for energy conservation and emission reduction. Meanwhile, in order to strengthen implementation in place, the Chinese Government set out the constraint objective of emission reduction by levels, and assigned goals for emission reduction to the relevant industries and to governments at all levels. On the other hand, the MEP has taken the regular scheduling and inspection system for emission reduction to supervise implementation of emission reduction goals, and implemented it in combination with the progress of emission reduction targets, adopting some important means, such as "regional limited approval", to establish the emission constraint mechanism. In addition, it has taken incentive measures, such as subsidies for desulphurisation electricity price to motivate enterprises to adopt emission reduction measures.

2.5.3 Environmental impact assessment

In October 2002, the 30th conference of the Ninth Session of the Standing Committee of the National People's Congress (NPC) deliberated and approved the Law of the People's Republic of China on Appraising Environmental Impacts, which establishes the EIA system for projects and planning. In August 2009, the State Council promulgated the Regulations on Environment Impact Assessment in Planning, marking a new stage of environmental protection integrated with comprehensive decision-making. The Regulations stipulated that the relevant departments under the State Council, local governments and their relevant departments above the municipal level with the planned districts shall conduct environmental impact assessments on the land-use planning, planning for building development and utilisation of the relevant regions, basins and sea areas (referred to as 'the comprehensive planning'), as well as the relevant special planning for industry, agriculture, animal husbandry, forestry, energy, water conservation, transportation, urban construction, tourism, natural resources development (referred to as 'the special planning'). Since the 11th Five-Year Plan, the number of EIAs at the planning stage is increasing year by year; the MEP has completed a total of 158 reviews of key regional planning EIAs, and local governments have completed more than 2 300 reviews of planning EIAs.

2.5.4 Pollution charges and emissions trading

In order to set the objectives for incentive and restraint mechanisms favourable to environmental protection, the Chinese Government has continuously improved its policies on pollution charges, further standardised the pollution charge system, and widened its coverage. By setting stricter collection criteria and strengthening collection and inspection, the environmental benefits of pollution charges are gradually increasing. Twelve provinces, including Jiangsu, Beijing and Tianjin, raised the bar for SO_2 emission charges and sewage charges, and regulated urban sewage charges. Since 2009, 10 out of 36 large and medium-sized cities included in the statistics raised the bar for sewage treatment charges substantially.

Since 2009, the Chinese Government has initiated emission-trading pilots in 18 provinces and cities, such as Jiangsu and Zhejiang, and some provinces and cities have established provincial emissions trading centres, such as Zhejiang, Beijing, Shanxi and Chongqing. In 2011, the Zhejiang Provincial Government issued the Rules of Zhejiang Province for Implementing the Interim Measures of the Paid Use of Emission Right and Trading Pilots to guide emission trading. Opinions of the State Council on Strengthening Environmental Protection explicitly proposed that China will set up a national emissions trading centre to boost emission-trading market during the 12th Five-Year Plan period.

2.5.5 Green credit and insurance

For green credit: the MEP, the People's Bank of China and the China Banking Regulatory Commission (CBRC) issued several joint policy papers, such as Opinions on Implementation of Environmental Policies and Laws & Regulations to Prevent Credit Risk and Circular on Fully Implementing the Green Credit Policy to further improve information sharing. More than 20 provinces and municipalities have promulgated the implementation files. Some provinces, such as Jiangsu and Guangdong, conducted credit evaluations on companies' environmental behaviour. The departments of environmental protection and financial institutions established a favourable information communication and sharing mechanism. About 40 000 items of information on companies' environmental violation, and more than 7 000 items of information on environment approval have been brought into the credit system of the People's Bank of China. In February 2012, the MEP issued the Green Credit Guideline to specifically regulate and guide green credit for the banking institutions in order to actively co-operate with the CBRC.

The State Council issued Administration Regulations on Taihu Lake in September 2011, which was the first time a liability insurance system in the form of administrative regulations of the State Council had been established. The MEP, together with the China Insurance Regulatory Commission (CIRC), issued Guidance on Liability Insurance for Environmental Pollution. Fourteen provinces (autonomous regions, municipalities directly under the central government), such as Shanghai and Chongqing, carried out pilots, and released their implementation opinions. Some provinces, such as Guangxi, Hunan, Hebei and Jiangsu, began to implement the pilots of compulsory liability insurance for environmental pollution for companies with high environmental risk. According to the incomplete statistics, more than 1 800 companies across the country had been insured, with premiums amounting to 102 million CNY, with insurance coverage reaching 17.43 billion CNY by the end of 2011.

2.5.6 Eco-compensation

In 2008, the revised Prevention and Control of Water Pollution of the People's Republic of China made specific provision on the eco-compensation mechanism for water environment in the form of law for the first time. The MEP issued Guidance on Implementation of Eco-compensation Pilot, and together with the MoF and MLR issued Guidance on Gradually Establishing Responsibility Mechanisms for Mine Environment Improvement and Ecological Restoration. Some provinces, such as Hebei, Henan, Jiangsu, Shandong, Liaoning, Shannxi and Sichuan, carried out eco-compensation pilots for key river basins, the important ecological function areas and mineral development. In order to further standardise eco-compensation mechanisms and promote eco-compensation works, eco-compensation legislation has already been listed in the agenda of the Chinese Government; Regulations on Eco-compensation have been listed as Category II items in the 2011 legislative scheme of the State Council, which will be led by the NDRC to formulate with jointly participating of the MEP and other relevant departments.

2.5.7 Science and technology of environmental protection

The Chinese Government highly values and constantly improves the technology to support environmental protection; Outline of the National Programme for Long- and Medium-Term Scientific and Technological Development (2006-2020) set up 16 major projects, four of which are related to environmental protection. During the 11th Five-Year Plan period, the MEP implemented three foundational and strategic environmental protection projects with major on the pollution source census, environmental macroscopic strategic research, and water pollution control and treatment technology as the main contents; learned about the national pollution situation and established an information database of pollution sources by conducting pollution source census; put forward the strategy and measures for future environmental protection in China through the national macroscopic strategic researches on the environment; constructed two preliminary technical systems of water pollution control and water environment management in China through implementing the national major water projects, which provided strong support to the national comprehensive improvement of water environment and security of drinking water. Meanwhile, the MEP drove technology platform construction, such as scientific research institutions and key laboratories, aggregated and cultivated scientific talent, and strengthened the research team for environmental protection through implementing major scientific research projects.

2.5.8 Environmental awareness and education

In order to improve public awareness of environmental protection and enhance the public capacity to participate in environmental policy-making, the Chinese Government used various media to carry out promotional activities for environmental protection. The MEP actively carries out various influential and creative theme activities on annual World Environment Day, so as to raise public awareness of environmental issues and encourage the public to participate actively in environmental protection. Since 2005, the MEP has released the Chinese theme of World Environment Day, correspondence to the theme of World Environment Day issued by the UNEP (United Nations Environment Programme) each year.

The Chinese Government responds in a timely manner to public concerns about environmental protection. In view of pressing issues by the media and netizens about above-standard blood-lead levels, air pollution, chromium slag dumping, hydropower development, and emission reduction policy, the MEP avails itself of various means, including news releases, press conferences, and exchange meetings by news spokesmen to address the issues with the responsible people from relevant departments and bureaus or invited experts, which have been favourably received by public opinion. In addition, it also encourages and supports the development of a number of environmental NGOs.

Through various activities, such as dialogue with NGOs, the MEP has always attached great importance to the role of grass-roots environmental organisations in undertaking environmental protection, and actively supports the role play of the NGOs. China has seen the emergence of a large number of influential domestic and foreign environmental NGO in recent years, such as Friends of Nature, Global Village, Institute of Public and Environmental Affairs, etc. Environmental NGOs also have an increasing influence in undertaking environmental protection in China. For example, Global Village has developed into a community with particular influence for the government, beyond its original role in community health maintenance and refuse classification; its main activities now include the establishment of green communities and the cultivation of ecological villages, etc..

The MEP actively carries out environmental professional education, organises and carries out the construction of green schools, green communities and green families, and carries out environmental-responsibility training in companies. A training class on company environmental liability has been held every year since 2010.

2.5.9 International co-operation and performance

By continuously raising the comprehensive national strength, and increasingly growing its international status and influence, international environmental co-operation in China has transformed from the closed to the open, and has gradually integrated into the international community. More and more international environmental issues need to be solved with the participation of China, which has progressed from "passively dealing with" to "active participation in" to "playing a constructive role".

The Chinese Government takes full advantage of the international co-operation platform for environmental protection, actively introduces advanced environmental concepts and management mechanisms, and learns from various experiences and lessons. It constantly improves the senior policy advisory mechanism of the China Council for International Co-operation on Environment and Development, improve the environment and development potential of China, and promotes the formation of China's environmental administrative system. Many significant environmental policies, systems, laws and regulations issued by the Chinese Government have referenced successful experiences and lessons from developed countries, such as the United States, Japan and Europe, avoiding the original approach of "firstly pollution, and then treatment" and "exchange environmental pollution for economic growth", as experienced by the developed countries, which contributes to the active exploration of a new path for China's environmental protection.

As for international conventions, the MEP has led the negotiations and performance of five international conventions on environment, two international conventions on nuclear safety, and four protocols. From 2002 to 2012, there have been a total of four new ratified conventions and protocols, including the Stockholm Convention on Persistent Organic Pollutants and the Biosafety Protocol. Over the past ten years, the MEP has been deeply committed to actively and effectively participating in international environmental conventions, and has made some fruitful achievements, which positively contributed to global sustainable development.

3. Development achievements

3.1 Develop circular economy with positive progress made in structural adjustment

Since the 11th Five-Year Plan period, the Chinese Government has already realised that comprehensive environment decisions play an important role in protecting resource environment and optimising economic growth. China is positively exploring a new path of industrialisation, adjusting industrial structure, developing a circular economy and promoting the development of strategic new industries, as well as transforming and upgrading its traditional industries as important ways to strive to transform the economic development pattern. It is promoting industrialisation by information, driving information by industrialisation, and boosting the manufacturing industry to improve core competitiveness, while continually accelerating the development of a modern service industry, actively advocating green consumption, and gradually improving development quality. Meanwhile, effective work mechanisms have been built among ministries, such as the NDRC, MEP, MLR, MWR (Ministry of Water Resources), MoA (Ministry of Agriculture), to co-ordinate the conflict between economic development and resource and environmental protection.

During the 11th Five-Year Plan period, energy consumption per unit of GDP in China has come down by 19.1%. Meanwhile, the Chinese Government gave full play to the reversed mechanism of pollution reduction, closed down small thermal power units of 76830 kW in total, accomplishing the close-down of 50000 kW a-year-and-a-half ahead of schedule; phased out the backward production:120 million tons of iron, 72 million tons of steel, 370 million tons of cement, 93 million tons of coke, 7.2 million tons of paper-making, 1.8 million tons of alcohol, 300 000 tons of gourmet powder (monosodium glutamate) and 38 million weight cases of glass (as shown in Figure 3.1).

In 2005, the State Council issued Opinions on Accelerating the Pace of Development of a Circular Economy, promulgated the policies related to finance, tax, investment and financing in a bid to effectively guide and support the development of a circular economy. In 2006, the key technologies of a circular economy were listed in the Outline of National Programme for Long- and Medium-Term Scientific and Technological Development. In 2008, the Circular Economy Promotion Law was issued, which is the third special law on circular economy in the world, following Germany and Japan. Since 2005, efforts have been made in organising and launching national circular economy demonstrations, and determining two batches, totalling 178 pilot units. Twenty-eight provinces (cities, counties) have launched provincial-level pilots. A total of 133 cities (districts, counties), 256 industrial parks and 1 325 companies have been chosen for pilots. Sixty typical pattern cases of circular economy with Chinese characteristics have emerged after summing-up and refining. In 2010, the output value from resource cyclic utilisation industry exceeded 1 trillion CNY, with over 20 million employees. Raw materials of one-fifth to one-third of products such as steel, non-ferrous metals, and paper pulp came from renewable resources; 20% of raw materials for cement came from solid waste; the comprehensive utilisation rate of industrial solid wastes reached 69%.

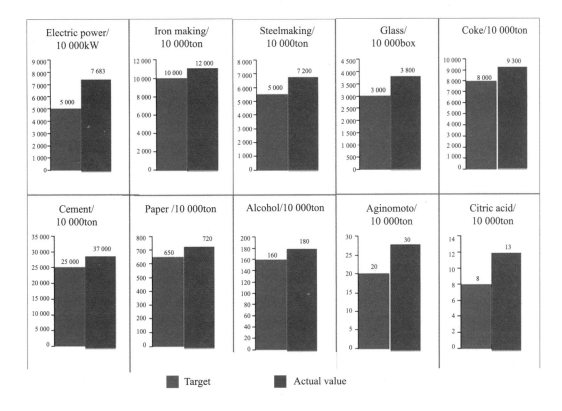

Figure 3.1 Situations of backward production capacity phased out during the 11th Five-Year Plan period

3.2 Emission of major pollutants significuntly reduced and remarkable outcomes have been achieved in key industries

During the 11th Five-Year Plan period, both economic growth and total energy consumption have exceed expectations, the SO_2 emission- reduction target was met a year ahead of schedule, and the urban and industrial COD emission target was met six months ahead of schedule. In 2010, total urban and industrial emissions of COD and SO_2 around the country have been reduced by 12.45% and 14.29% respectively compared with 2005 Level. Both have accomplished an extra 10% of the targeted value (see Figs. 3-2, 3-3)with 2005 level. Both In 2010, COD emission intensities in the paper-making, chemical and textile industries have been reduced by 73.9%, 66.7% and 50% respectively compared with those in 2005. SO_2 emission intensities in the electric power, non-metallic mineral product, and ferrous metal smelting sectors have been reduced by 72.5%, 58.1% and 50%, respectively.

SO_2 removal rates in important industries have steadily improved, with remarkable effects on SO_2 emission reduction. In 2010, the industrial SO_2 removal rate was 66%, up 30.5% over 2005, in which SO_2 removal rates in electric power production, ferrous metal metallurgy, chemical industry, non-ferrous metallurgy, and petrochemical industry reached 68.3%, 31.9%, 54.2%, 89.9% and 79% respectively, which was 50%, 8.5%, 5.5%, 3.9%, and 23.3% higher respectively than those in 2005 (see Figure 3.4). During the 11th Five-Year Plan period, the reduction ratio of SO_2 emission in the electric power industry, which accounts for over half of industrial SO_2 emissions, was 22.9%, higher than the ratio of total SO_2 emission reduction.

The COD removal rate of wastewater in key industries has been continuously improved; COD emission reduction results have been remarkable. In 2010, the industrial COD removal rate was 80.2%, up 10.1% over 2005, of which COD removal rates in paper-making, the chemical industry, and the farm-products processing industry respectively were 82.4%, 76% and 76.7%, an increase of respectively 15.2%, 21.7% and 38.5% from those in 2005 (as shown in Figure 3.5); during the 11th Five-Year Plan period, the ratio of COD emission reduction in the paper-making industry was 40.4%, which was higher than the ratio of total COD emission reduction.

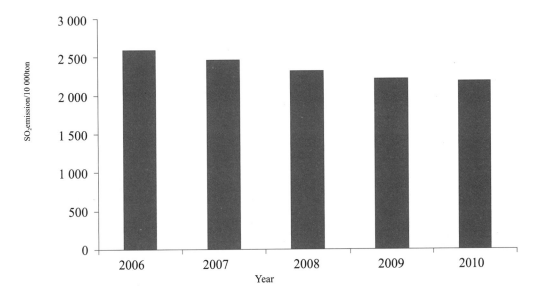

Figure 3.2 SO_2 emissions during the 11th Five-Year Plan period

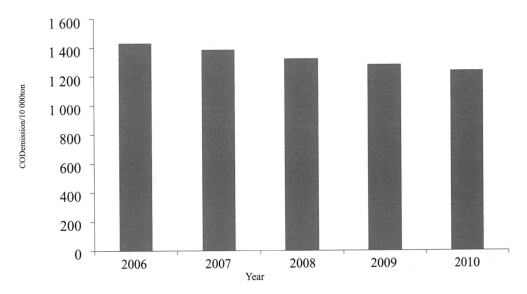

Figure 3.3 COD emissions during the 11th Five-Year Plan period

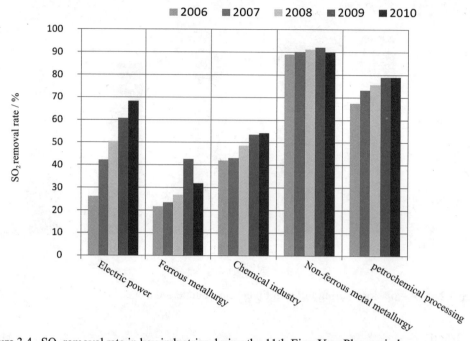

Figure 3.4 SO$_2$ removal rate in key industries during the 11th Five-Year Plan period

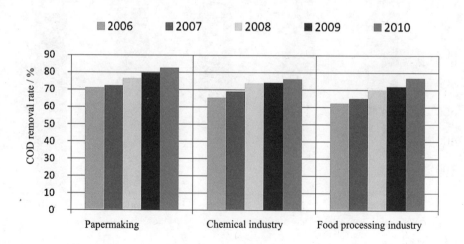

Figure 3.5 Variation of COD removal rate in key industries during the 11th Five-Year Plan period

3.3 Strengthened pollution control in key basins and regions with improved environmental quality in local regions

By implementing a total emission-control system of major atmospheric pollutants, efforts have been made to strengthen industrial pollution source control, improve the consumption ratio of urban clean energy and energy utilisation efficiency, and intensify the adoption of certain measures, such as pollution control of vehicles, which has resulted in improvements in atmospheric environmental quality. During the 11th Five-Year Plan period, the ratio that reached the second-grade standard of China's urban air quality has been rising year by year, while the city ratio that reached the third grade or below the third grade was on a downward trend. The percentage of cities that reached the second-grade standard rose from 58.1% in 2006 to 79.2% in 2010, while the percentage of cities that reached the third-grade standard decreased from 28.5% in 2006 to 15.5% in 2010. The percentage of cities that are below the third-grade standard has decreased from 9.1% in 2006 to 1.7% in 2010 (as shown in Figure 3.6). Annual average concentrations of SO_2 and PM_{10} in national urban environmental air have been reduced by 26.3% and 12% respectively (as shown in Figure 3.7).

In 2010, among the State-controlled monitoring sections of surface water, the ratio of sections with Grade I-III water quality is 14.4% higher than it was in 2005, while the ratio of sections with worse than Grade V water quality was reduced by 6.6 %. The degree of water eutrophication in state-controlled key lakes and reservoirs has been gradually mitigated. In 2010, among 26 state-controlled key lakes (reservoirs), one reached Grade II water quality, accounting for 3.8%; five reached Grade III water quality, accounting for 19.2%; four reached Grade IV water quality, accounting for 15.4%; six reached Grade V water quality, accounting for 23.1%; and 10 reached worse than Grade V water quality, accounting for 38.5%. From 2005 to 2010, annual COD concentration of state-controlled monitoring sections of surface water in 113 environmental protection cities dropped from 7.2mg/L to 4.9mg/L, and the percentage of the drop is 31.9% (as shown in Figure 3.8).

During the 11th Five-Year Plan period, the intensity of severe acid rain has been declined. There has been a declining trend in cities that suffered less severe acid rain (annual rainfall pH<5.0) and severe acid rain (annual rainfall pH<4.5). The ratios of cities with less severe and severe acid rain have dropped from 10.7% and 17.7% in 2006 to 8.5% and 13.1% in 2010 respectively.

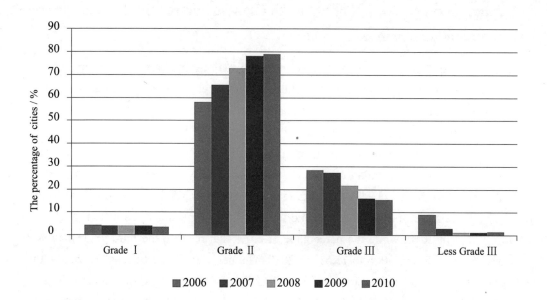

Figure 3.6 National air quality level

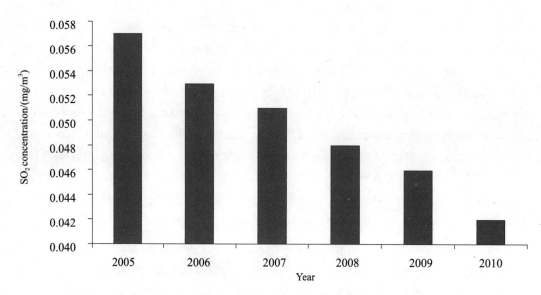

Figure 3.7 SO_2 concentration in 113 key environmental protection cities

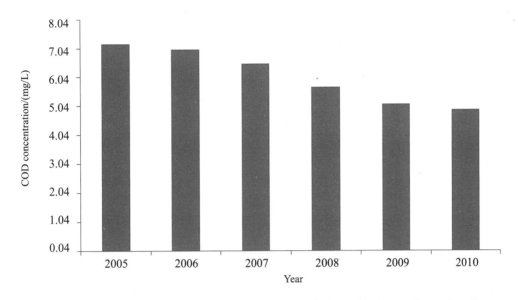

Figure 3.8　COD concentration change of 759 state-controlled monitoring sections of surface water

The result of pollution control over key basins and regions was remarkable. According to the requirements stipulated in the Interim Evaluation Measures for Implementation of Special Pollution Control Programmes in Key River Basins and Regions, an examination system of the water quality of provincial boundary sections in key river basins and regions has been fully established. The ratio of pollution control for special planning projects in key river basins and regions has reached 87%, which is 22.8% higher than that during the 10th Five-Year Plan period, with completed investment up to CNY 138.9 billion in total. Following the success of the "Green Olympics", the tasks of air quality guarantee for Shanghai Expo and Guangzhou Asian Games have been successfully accomplished.

3.4 Comprehensively promote ecological protection with remarkable outcomes in ecological construction achieved

As a signatory country of the Convention on Biological Diversity, the Chinese Government attaches great importance to biodiversity protection. It issued and implemented an Outline of National Programme for Biological Species Resource Protection and Utilisation (2006-2020)

in 2007, issued and implemented Chinese Strategy and Action Plan for Biological Diversity Protection (2011-2030), as well as other plans and programmes in 2010, proposing 3 stage goals, 10 priority fields, 30 priority actions and 39 priority projects in China's biodiversity protection, which have become the programmatic documents for the country's biodiversity protection. In 2010, the State Council set up the National Commission on China Biological Diversity Protection with Vice Premier Li Keqiang acting as its president, and organised the joint inter-ministerial meeting of biological species resource protection.

In order to comprehend the conditions of Chinese important biological species resources and biodiversity, the MEP launched a national investigation on important biological species resources from 2004 to 2009, together with relevant departments; it completed the catalogue and investigation report on relevant species, established species distribution and protection of key wildlife resources, livestock and poultry breeds resources, drug organism resources, and microbial strain resources, and established a national biological species resource database and information platform in China. From 2007 to 2012, national biodiversity assessment was organised and implemented. By the end of 2011, all provinces and cities had finished biological diversity assessment. These assessments provided a preliminary understanding of the current conditions of biodiversity, space distribution features and main threats in all provinces and cities throughout the country. The county-based biodiversity assessment databases have been arranged and formed, and counter-measures and suggestions on biodiversity protection and sustainable utilisation have been put forward for various provinces.

With regard to natural reserve construction, the number of natural reserves across the country have increased from 2 531 to 2 640 from 2007 to 2011 (see Figure 3.9). The total area of natural reserves in 2011 was about 1.49 million square kilometres (of which about 1.43 million square kilometres were terrestrial areas, and about 60 000 square kilometres marine areas), with the terrestrial area of natural reserves making up about 14.9% of China's total land area.

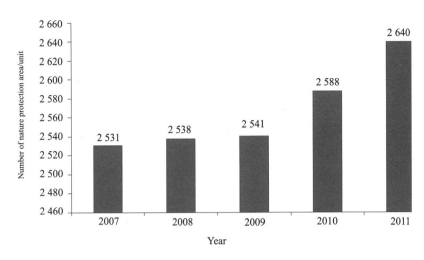

Figure 3.9 Growth in number of natural reserves

Protection capacities of woodlands, grasslands and wetlands have been steadily strengthened and there has been a net reduction of desertification land area. As regards woodland protection, in recent years the forestry departments at all levels in China have made great efforts in planting trees on the barren mountains and their surroundings, carrying out the construction of characteristic economic forests, carbon-sink forest and energy forest bases, strengthening the cultivation of high-quality native soil and rare tree species. From 2007 to 2010, the total area of forestation in China increased from 3 907 711 hectares to 5 909 919 hectares with growth of 51.2% (see Figure 3.10). In terms of grassland ecological protection, the total grazing-banned grassland area reached 80 667 000 hectares in all provinces and regions; grass-livestock balanced grassland areas reached 170 667 million hectares; 1 987 000 husbandry households enjoyed subsidies for capital goods by the end of 2011. With regard to wetland protection, since 2009, China launched the second national wetland resource survey, preliminarily setting up the wetland-protection network system mainly based on wetland natural reserves and wetland parks. In the meantime, China has been seriously applying the Convention on Wetlands, strengthening its role in carrying out such obligations stipulated in the Convention, successfully implementing the international co-operation programme by working together with Australia, Germany and the United States, etc. Up to 2011, China had increased wetland protection areas by 330 000 hectares, restored 23 000 hectares of wetlands, and added four international key wetlands and 68 national wetland park pilot sites. The number of international key wetlands have reached 41, covering 3 710 000 hectares of area, while wetland demonstration zones have reached 3 490 000 hectares. In terms of desert ecosystem protection and treatment, in recent years China has actively pushed forward the construction of state-level desertification forbidden reserves and regional sand prevention and control, and completed the fourth desertification and land desertification monitoring. During the 11th Five-Year Plan period, the Chinese Government completed desertification treatment of 10.81 million hectares. The area of desertified land has decreased from the annual average expanded area of 3 436 square kilometres in the last century to 1 717 square kilometres on an annual average basis today. A net reduction of desertified land has been achieved on the whole.

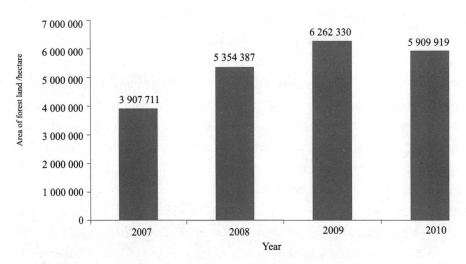

Figure 3.10 Total area of planted forests

3.5 Remarkable achievements has made capacity building of Environmental-protection and enhancing environmental-regulation capacity

Environmental infrastructure construction has made progress by leaps and bounds. During the 11th Five-Year Plan period, a total of 578 GW coal-fired power plants with desulphurisation facilities were built and put into operation. The ratio of thermal power desulphurisation units rose to 82.6% from 12% in 2005; 2 832 urban centralised sewage treatment facilities were built, with daily treatment capacity up to 125 million cubic meters. During the 11th Five-Year Plan period, 2 000 urban centralised sewage treatment facilities were built, with additional sewage treatment capacity up to 50 million tons per day. The national urban sewage treatment rate rose to more than 75% from 52% in 2005. Sewage treatment plants were built in all the counties in some provinces, such as Henan, Jiangsu, Zhejiang and Guangdong. Ningxia was the first province in North-West Region to launch the building of sewage treatment plants in all counties.

By virtue of capacity building for environmental regulation and funding for environmental protection work, an environmental regulating system based on supervision by the state, regulated by local governments and organised by relevant units was established and perfected. Across the country, 52% of county or district-level environmental monitoring stations have equipped sufficient devices to fulfill the standardisation. Environmental supervision and monitoring enforcement has steadily improved. Since 2006, special inspections of heavy-metal pollution and paper-making enterprises, sewage treatment plants, waste landfills, etc. have been carried out. More than 10.65 million enforcement personnel took part in the inspection of over 4.46 million companies, of which more than 80 000 found to be violating environmental laws were investigated and punished, 7 239 discharging pollutant illegally were banned and shut down, 5 981 were required to suspend production for rectification, 6 432 were required to perform treatment within a limited period, and 19 000 cases against environmental protection law were handled under the supervision of a higher authority. This campaign has severely cracked down on environmental violations and safeguarded the public's environmental rights and interests.

Environmental emergency response systems and mechanisms have been increasingly perfected. The MEP has set up an emergency centre and built an environmental emergency expert base. More than one-third of provincial environmental departments have established special environmental emergency management agencies. The department emergency linkage mechanism has been strengthened. According to the requirements stipulated in the Circular on Establishing and Improving the Emergency Linkage Mechanism for Environmental Protection and Safety Supervision Departments issued by the MEP and State Administration of Work Safety, the environmental departments and safety supervision departments of nearly 20 provinces (autonomous regions, municipalities directly under the Central Government) have signed emergency linkage mechanism agreements at present, which has further strengthened the environmental protection and safety supervision emergency incidents linkage. During the 11th Five-Year Plan period, a total of 912 environmental emergencies incidents have been dealt with, and a series of key or serious environmental emergencies of high concern to the public have been properly handled.

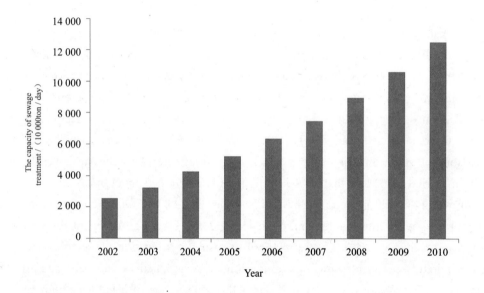

Figure 3.11 Urban sewage treatment capacity

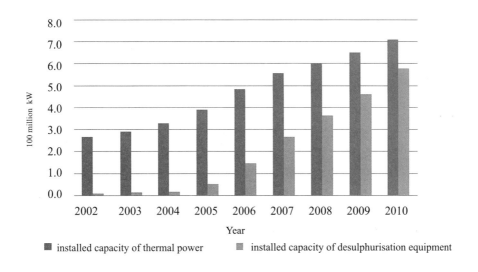

Figure 3.12 Installed capacity of thermal power plants and desulphurisation equipment

3.6 Information disclosure work performed in a more normative way and extent of public participation continuously strengthened

On May 1st 2008, the former State Environmental Protection Administration issued and implemented the"Regulations on the Disclosure of Government Information and Measures on Environmental Information Disclosure",which has regulated and promoted environmental information disclosure by governments and enterprises. The Chinese Government has more strengthen on environment supervision and environmental information disclosure. With regard to the information disclosure, the MEP has issued an environmental statistics yearbook and environmental quality bulletin annually. Meanwhile, it publishes pollution-source supervision information on the homepage of the MEP website. Environmental protection departments at all levels also release information related to environmental quality through quality bulletins, media and other ways.

China's environmental-protection NGOs actively participate in policy and decision-making for the joint promotion of the development of environmental-protection undertakings. In recent years, several environmental-protection NGOs, such as Friends of Nature, Global Village, and the Institute of Pubic and Environmental, which are influential both at home and abroad, have emerged in China. The Chinese Government encourages all stakeholders, including experts and departments in NGOs, to participate in environmental decision-making, and establish interactive channels interfacing with the public by media, network and communications, etc. For example, for the purpose of strengthening the scientific nature of plan preparation and improving the implementation effect of plans, the MEP has emphasised the participation of the public in plan preparation for key river basins and regional pollution control. By adopting internet voting, consultation meetings, planning information disclosure and other approaches, the participation of stakeholders has been highlighted. Moreover, active feedback has been given to the supervisory opinions proposed by NGO during plan implementation.

4. Conclusions and prospects

Following the Industrial Revolution and Information Revolution, the "Green Revolution" has become a catalyst for a new round of global economic transformation. Especially since the global economic crisis of 2008, people have become soberly aware that the "black" economic development mode in resource depletion and relying on fossil fuels is unsustainable. Human beings must seek a green way. As an emerging economic power, the green economy will have a far-reaching influence on China's future economic prosperity as well as the global economy. The Chinese Government realised that the pressure on resources and the environment faced by China may be worse than in any other country in the world. Environment and resource issues are more prominent than in any other country, and it is also more difficult for China to solve them than any other countries. Transforming the economic development mode and achieving green development is therefore not only a strategic choice and inevitable need of the Chinese nation for long-term development, but also a positive contribution to global sustainable development, which will have an important influence on human development.

Since the 11th Five-Year Plan, great changes are taking place in China's environmental protection from awareness to practice. Environmental protection is regarded not only as an effective mean to optimise economic growth and force a green transformation of the economic growth mode, but also as the important starting point to enhance the economic competition advantage and hold a favourable position in the future global economy. It has become a consensus of both governments at all levels and the public. Moreover, through the implementation of the 11th Five-Year Plan, the Chinese Government has completed each constraint target and index in the Plan, meanwhile strengthening the environmental regulatory capacities of governments, and promoting the development of an environmental-protection industry and technology, which provides a basis for the 12th Five-Year Plan and longer-term economic growth.

During the 11th Five-Year Plan period, the achievements of China's environmental policies depended on the following three factors: the first is the establishment of powerful implementation mechanisms for environmental planning and policies through improving the system mechanism. It is guaranteed that all environmental indexes, including total emissions reduction should be carry out from the central to the local levels through the target responsibility system and effective performance evaluation mechanism. In addition, for the environmental field, the Central Government strengthened the supervision of environmental planning and policy implementation by setting up the MEP and a number of regional supervision centres; the second is the implementation of a series of effective policy measures. In the field of energy conservation and emissions reduction, the Chinese Government not only emphasised strengthening the constraints of policy objectives to governments at all levels and to industries through the law and administrative means, but also paid attention to economic means; for example, the policy of desulphurisation electricity price subsidies positively promotes emissions reduction in electric power companies. The third is to provide the basic guarantee for implementing policies and measures through a series of capacity building. For the field of environmental monitoring, the Chinese Government invested a total of 30 billion CNY over five years to strengthen capacity building in environmental monitoring, environmental supervision, environmental emergency response, environmental statistics and environmental enforcement.

However, China will face continuous environmental pressure on resources. Though due to the global financial and economic crisis since the end of the 11th Five-Year Plan, China's economy has kept on a growth trend, with the rapidly increasing demand for energy resources being maintained. However, the transformation of economic structure and growth mode will still face greater challenges. Meanwhile, due to long-standing historical causes, the deterioration trend in China's overall ecological environment has not been fundamentally saved with environmental risks still in a state of frequent occurrence. The current complex resource and environmental problems will be solved not only on the basis of speeding up the adjustment of the existing economic growth mode and structure, but also fully mobilising strengths, including the international community, to participate in and support China's green development process.

In the 12th Five-Year Plan, the Chinese Government established more inclusive development objectives for green development. The overall objective of China's 12th Five-Year Plan is to accelerate adjustment of the economic growth mode and structure. In order to achieve an inclusive, green and competitive economic development mode, the Plan contained a total of eight macro-economic and environmental development indexes which are directly related to the development of a green economy. For the harmonious development of the overall population, resources and environment, the Chinese Government proposed to implement the comprehensive regional development strategy and main function regional strategy, build a regional development pattern with features of complementary advantages for regional economies, adopt a clear main function orientation, efficient utilisation of land and space, harmonious co-existence between human beings and nature, and gradually achieve equal access to basic public services for the different areas, so that more development achievements can benefit the widest public.

ANNEX:

ACTIONS TAKEN SO FAR AFTER THE 2007 OECD'S ENVIRONMENTAL PERFORMANCE REVIEW OF CHINA

No.1

Recommendations

Implement environmental law and regulations nationwide for products and industrial/energy facilities; strengthen monitoring, inspection and enforcement capabilities throughout the country, including through the independence of the enforcement functions of Environmental Protection Bureaus (EPBs).

Implementation

At present, there are more than 50 laws and regulations on energy/resource management and environmental protection issued in China, including the Constitution of the People's Republic of China, Environmental Protection Law of the People's Republic of China, Electricity Law of the People's Republic of China, Coal Industry Law of the People's Republic of China, Energy Conservation Law of the People's Republic of China, Renewable Energy Law of the People's Republic of China, Mineral Resources Law of the People's Republic of China, Water Pollution Prevention and Control Law of People's Republic of China, Atmospheric Pollution Prevention and Control Law of the People's Republic of China, Cleaner Production Promotion Law of the People's Republic of China, Circular Economy Promotion Law of the People's Republic of China, and Law of the People's Republic of China on Prevention of Environmental Mental Pollution Caused by Solid Waste.

Since the 11th Five-Year Plan, the Chinese government has constantly improved environmental laws and regulations and relevant standards on the industrial pollution and energy field; it revised and issued Renewable Energy Law of the People's Republic of China (amendment); formulated and implemented Circular Economy Promotion Law of the People's Republic of China; and successively promulgated seven administrative regulations, such as the Regulation on Planning Environmental Impact Assessment, and the Administrative Regulations on Recycling and Disposing Electrical and Electronics Waste. And meanwhile, more than 60 standards on pollution emissions of key industries have been revised, and revision of 1 050 national standards on environmental protection have been carried out. At present, the total quantity of the national standards on environmental protection is 1 300, which doubled the number in the 10th Five-Year Plan. These legislation and standards provide a fundamental basis and guarantee to strengthen pollution control in the industrial field, reduce pollution emissions per product, and improve energy efficiency as well.

During the 11th Five-Year Plan Period, based on implementation of energy conservation and emission reduction policy, the Chinese government fully strengthened the capacity building for environmental regulatory, formulated and implemented the 11th Five-Year Plan on National Capacity Building for Environmental Regulatory; at present, the relevant investment of 30 billion CNY has been completed. With efforts over five years, capacity of environmental regulation has been improved significantly.

NO.2

Recommendations

Consider establishing SEPA as a ministry; strengthen SEPA's supervisory capacity of EPBs in local government.

Implementation

Concerning institutional reform, the MEP (Ministry of Environmental Protection) was formally established in March 2008, successively adding Department of Total Pollutants Control, Department of Environmental Monitoring, Department of Education and Communication, Department of Nuclear Safety Management (Department of Radioactive Safety Management), and Centre of Satellite Environment Application, etc. The newly established MEP strengthens overall planning and co-ordination of major issues, such as total pollutants control, environment supervision, nuclear and radiation supervision, with the establishment of MEP, each province has upgraded the provincial Environmental Protection Bureau to the provincial Department of Environmental Protection.

Since 2006, the former SEPA has formed six regional supervision centres for environmental protection, including North China, East China, South China, Northwest, Southwest, and Northeast, as well as six regional supervision stations of nuclear and radiation safety, including the Northern, Shanghai, Guangdong, Sichuan, Northeast and Northwest; among of them, the former is the agencies of the MEP, whose main functions are to supervise the local implementation of national environmental policies, planning, regulations and standards; investigate the cases of serious environment pollution and ecological destruction, co-ordinate and deal with the significant trans-provincial, basin and coastal environmental disputes. Some provinces and cities have also established independent regional supervision centres; for example: Jiangsu Province has formed three regional supervision centres of environmental protection for south Jiangsu, middle Jiangsu and north Jiangsu separately; Shannxi Province has also established the North Shannxi Supervision Centre of Environmental Protection.

Through organisational restructuring, the management system of environmental protection has been further smoothed, the environmental monitoring system of "national supervision, local regulatory and entity be responsible" is set up.

NO.3

Recommendations

Continue efforts to make local leaders more accountable to the higher level government and to local populations for their environmental performance.

Implementation

Concerning government performance evaluation, the Ministry of Supervision has organised and carried out government performance evaluation since 2008. 24 provinces (autonomous regions and municipalities) and more than 20 ministries under the State Council have carried out government performance management so far. Government performance assessment mainly focused on the completion of key work/tasks and the treatment of social concerns; the adoption of refined, visualised and quantised management index to make responsibilities more clear and process control more effective; gave full play of the guidance, incentive and restraint effect in performance management. At the same time, environmental performance evaluation has also been carried out on some local governments. For example:

through two environmental management systems of "the quantitative assessment on comprehensive improvement of the urban environment" and "creating the national model city of environmental protection", the management mechanism for urban environment that "led by the local governments, divided the responsibility of each departments, and unified supervision and management by the environmental protection department, and actively participated in by the public" was gradually established, which has played a positive role in urging the local governments to increase input in environmental protection and promote sustainable development.

Environmental protection and supervision responsibilities of local governments have been redefined. The current Environmental Protection Law of the People's Republic of China states:"the local people's governments at various levels shall be responsible for the environment quality of areas under their jurisdiction and take the necessary measures to improve it". In February 2008, the National People's Congress revised the Water Pollution of Prevention and Control of People's Republic of China, which declined to take the completion of water environment protection target as evaluation content for local governments and principals, and strengthened the responsibility of the local government on water pollution prevention and control. In addition, the modification of Environmental Protection Law of the People's Republic of China has been listed in the legislation planning of the 11th Standing Committee of National People's Congress; it is one of the key points to strengthen the environmental quality responsibility in this modification. In addition to environmental quality improvement, governments at all levels shall also be responsible for environment risk prevention, and emission reduction of the major pollutants.

As total emission reduction is one of the important environmental responsibilities undertaken by governments at all levels, the Chinese government has established a strict supervision system. Focussing on the total emission reduction goals, the Chinese government established the mechanism for goal decomposition, evaluation and rewards and punishment. The MEP, on behalf of the State Council, decomposed the emission reduction goals and assigned them to each province, autonomous region, and municipality directly under the central government, as well as eight central enterprises, including Xinjiang Production and Construction Corps and CNCP (China National Petroleum Corporation), SINOPEC (China Petrochemical Corporation), SGCC (State Grid Corporation of China), China Hua Neng Group, China Datang Group, China Huadian Corporation, China Guodian Corporation, and China Power Investment Group, the emission reduction goals are decomposed and implemented through signing a letter of target responsibility. The provincial government will further decompose and assign the target downward to form a goal decomposition and assessment mechanism from top to bottom. So it asserts great pressure on local governments which haven't initiated total emission reduction goals.

During the 11th Five-Year Plan period, a total of 111 local governments or enterprises have received punishment from the Central government, 6 regional governments and 4 Group Companies have been restricted for approval for the local regions, 26 mayors have been questioned by the Central Government, and more than 100 people received the administrative punishment. At the same time, the whole social pollution control ability and government environmental supervision capability were greatly enhanced and the environment quality was improved accordingly. During the 11th Five-Year Plan period, about 2000 sewage treatment plants were added, the capacity of coal-fired power plants with desulfurisation facilities which have put in operation increased for 532 million kW; percentage of thermal power desulfurisation unit was increased from 12% in 2005 to 82.6% in 2010; during the same period, sulfur dioxide concentration of 113 key environment-friendly cities fell by 26.3%; COD concentration at 759 surface water national monitoring section fell by 31.9%; so the percentage of Grade-II air quality in the prefecture-level cities is raised considerably.

Concerning environmental supervision, Opinions of the State Council on Key Works to Strengthen the Environmental Protection was promulgated in 2011 and provided "to strengthen the supervision on environmental enforcement; execute the supervision systems of the basin, regional, industry limit approval and the listed supervision, etc. the leadership of local governments who shall be responsible for unfinished environmental protection task or occurrence of great or abrupt environment affairs shall be questioned, and rectification measures shall be implemented in place." For example, the regional limited approval shall be implemented for the regions with frequent occurrence of serve and abrupt environmental accidents or prominent environmental risks and potential hazards; In 2012, the MEP prepared the Measures for Investigation on the Abrupt Environmental Accident and Responsibility. For the severe or abrupt environment accidents which were dispatched and disposed by the MEP since 2011, the associated local government and relevant departments which failed to perform the supervision responsibility of environmental safety were all urged to investigate the relevant administrative accountabilities.

In order to impel the local government to take more environmental responsibility, the resource consumption and environmental protection have been brought into the current achievements appraisal contents for the party and government officials as the important content of evaluation. In 2006, the Trial Implementation Measures of Comprehensive Evaluation on the Local Party and Governmental Leading Group and Officials with Embodying the Scientific Outlook on Development was issued by the Communist Party Central Committee's Organisation Department, which further improved and perfected the official evaluation; the assessment contents mainly include: the local GDP per capita and growth rate, revenue and growth rate per capita, the urban and rural resident incomes and growth rate, resource consumption and safety production, elementary education, urban employment, social security, land resources protection, environmental protection, science and technology input and innovation, etc..

NO.4

Recommendations

Strengthen the integrated permitting system and establish it as a more central instrument for pollution prevention and control; strengthen the integration of environmental protection in land-use planning and regulations, as well as in other relevant plans and regulations.

Implementation

The MEP actively made preparations for the legislation on discharge permit and promoted the legislation process. In 2007, the former SEPA formulated the Administrative Regulations on Discharge License (draft). In 2008, the new revision of Water Pollution Prevention and Control Law of the People's Republic of China explicitly stipulated discharge license as an important means to strengthen supervision of pollutants discharge, and defined the legitimacy position of water pollutant discharge permit system, so as to provide a legal basis for the implementation of discharge permits system. Since the 11th Five-Year Plan, the total emission reduction of major pollutants in whole nation also laid a foundation for legislation on discharge permits. In addition, China is also actively carrying out the pilot of comprehensive discharge license management. Since 2004, the former SEPA began carrying out pilot work of discharge license in Tangshan City in Hebei Province, Shenyang City in Liaoning Province, Hangzhou City in Zhejiang Province, Wuhan City in Hubei Province, Shenzhen City in Guangdong Province, and Yinchuan in Ningxia Hui Autonomous Region.

In order to comprehensively consider economic development and environmental protection, the Chinese government has incorporated the environmental protection into the related planning and regulations. In addition, Regulations on Environmental Impact Assessment of the Planning (Decree No. 559 of the State Council) issued and implemented in 2009 stipulated that "the planning related to land use, the planning related to construction, development and utilisation of region, basin and waters, as well as the special planning related to industry, agriculture, animal husbandry, forestry, energy, water conservancy, transportation, urban construction, tourism, and development of natural resources, shall conduct the relevant Environmental Impact Assessment". Outline of Overall Planning on National Land Use (2006-2020) was issued by China's State Council in 2008, which specified in Chapter 5 that: "the land use and ecological construction shall be co-ordinated with each other". The revision of Land Administration Law of the People's Republic of China has been listed in the legislation planning of the 11th National People's Congress Standing Committee, and will be organised to draft by the State Council.

Concerning development of national land and space, in order to form a co-ordinated development pattern among population, economy, resources and environment, the Chinese Government promulgated the National Planning for Major Function Zones in 2010, which proposed that China's land and space is divided into optimisation development zone, key development zone, limited development zone and prohibited development zone according to the development way; among them, optimisation development zone is mainly to speed up the transformation of economic development mode, adjust and optimise the economic structure, participate in the global division of labor and promote competition level; Key development zone is mainly to promote the sustainable economic development, promote its new industrialisation process and enhance the industrial cluster capacity on the basis of optimizing structure, improving efficiency, reducing the consumption of resources and protecting environment; Limited development zone mainly refers to the major producing area of agricultural products, where focuses on protecting the cultivated land, cultivating modern agriculture, enhancing the comprehensive agricultural production capacity, and increasing the farmers' income; Prohibited development zone refers to the key ecological function areas, which needs to limit large-scale and intensive industrialisation and urbanisation during the national spatial development, so as to maintain and improve the capacity of ecological products provision. The Planning also required adjusting and improving the related program, policy, laws and regulations on the finance, investment, industry, land, agriculture, population, and environment, etc., establishing and improving the performance evaluation system.

In addition, according to the Opinions of the State Council on Strengthening Environmental Protection (GF [2011] No. 35), the state will compile regionalisation of environmental functions, and define the ecological bottom line for the important ecological function areas, land and marine ecological sensitive area, and fragile areas, formulate the corresponding environmental standards and policies for all kinds of major function zones respectively; and strengthen the ecological environmental protection for ecological barrier of Tibet Plateau, Loess Plateau - Sichuan-Yunnan, Northeast Forest Belt, northern sand-prevention belt and south hill /mountainous region, as well as large rivers and important water systems.

NO.5

Recommendations

Extend the use of pollution charges, user charges, emissions trading and other market-based instruments and their incentive functions, taking social factors into account.

Implementation

Since the 11th Five-Year Plan period, China has continued to accelerate the development and implementation of a series of environmental and economic policies, and each local government also developed some environmental and economic policies according to their local situations; for example, the Regulations on Collection and Utilisation of Pollutant Discharge Fee stipulated that: the pricing competent department under the State Council, financial departments, the administrative departments of environmental protection and departments of economic and trade shall formulate the state criteria for collecting the pollutant discharge fee according to the development needs for the industrialisation of pollution control, requirements of pollution control, the economic and technological conditions, and bearing capacity of polluters. The provinces, autonomous regions, and municipalities directly under the central government which are not specified in the state criteria for pollution discharge fee collection, may prepare their local criteria for collecting the pollutant discharge fee. The Chinese Government has promoted energy conservation and emission reduction, environmental protection and the industrial restructuring, and achieved certain positive results through effectively using economic means.

The standardisation of the pollution charges system is being continuously improved, its coverage is gradually expanding, the collection criteria becomes gradually stricter; the collection audit is continuously strengthened; the total amount of pollution charge is increasing continuously; and the due role of pollution charges is also strengthened step by step. For example, 12 provinces, such as Jiangsu, Beijing and Tianjin enhanced the criteria for SO_2 emission charges. Besides, since China's criteria for sewage treatment fee is relative lower, the Chinese Government has proposed a reform thought of "break-even and meager profit" for urban facilities of water supply and sewage treatment. Since 2009, 10 cities among 36 large and medium-sized cities have sewage treatment fee with average increase rate of 29%, nearly 70% for a few cities. The IT construction for pollutant discharge fee collection has been also gradually valued, information disclosure means are applied to encourage enterprises to timely pay pollutant discharge fees; such as the public announcement released to inform the list of enterprises who failed to pay pollutant discharge fee and imposed a fine.

Emissions trading: since 2009, the MEP in conjunction with MoF (Ministry of Finance) have initiated the paid use of emission right and trading pilots in some provinces and cities, such as Zhejiang, Shannxi, Hunan, Hubei, Shanxi, Inner Mongolia, Hebei, Chongqing, and Henan. Zhejiang Provincial Government issued the Rules of Zhejiang Province for Implementing the Interim Measures of the Paid Use of Emission Right and Trading Pilots in 2011 to guide the emissions trading. Opinions of the State Council on Strengthening the Environmental Protection explicitly proposed that China shall set up a national emissions trading centre to cultivate emission trading market during the 12th Five-Year Plan period. At present, some provinces and cities in China have established the provincial emissions trading centre, such as, Zhejiang, Beijing, Shanxi and Chongqing.

Environmental tax: China has not established a separate environmental tax yet; however the related work is under progress. For example, the Chinese Government has promulgated the relevant policy papers, such as Directory of the Enterprises Income Tax Preference for Environmental Protection, Energy-Saving and Water-Saving Projects (for Trial), Directory of the Enterprises Income Tax Preference for Comprehensive Utilisation of Resources (2008), and Circular on Some Issues about Investment in Special Equipment for Environmental Protection, Energy-Saving or Water-Conservation, Production Safety to Credit the Enterprise Income Tax, which can support the environment-friendly project and equipment through tax policies, such as the income tax.

Pricing policy: the desulfurisation electricity pricing policy implemented in China speeds up construction and operation of the desulfurisation facilities in the thermal power industry, significantly promoted SO_2 emission reduction, and meanwhile, the Chinese Government also strengthens the difference electricity pricing policy for enterprises with high energy consumption and implements punitive electricity price, difference electricity price or stepped electricity pricing policy for the ultra energy consumption products. During the 11th Five-Year Plan period, the desulfurisation electricity price of 1.5 *fen*/kWh has been implemented; the Central Finance has subsidised CNY 400 000/km for the supporting construction of pipeline network of sewage treatment plant. In addition, the charge criteria are raised in most provinces for sewage treatment and garbage disposal in order to encourage enterprises to protect the environment. In November 2011, the pilots for denitration electricity price subsidies were began to carry out in 14 provinces in order to promote NO_x emission reduction.

Green credit: the MEP, the People's Bank of China and China Banking Regulatory Commission (CBRC) etc. have jointly issued a series of policy papers in recent years, such as Opinions on Implementation of Environmental Policies and Laws and Regulations to Prevent Credit Risk, Circular on Fully Implementing the Green Credit Policy to Further Improve Information Sharing, Guidance on Further Providing the Financial Services to Support the Restructuring and Revitalisation of Key Industries and Restrain the Surplus Capacity in Some Industries, and explicitly required the financial institutions strictly control credit to the outdated production capacity. In addition, Opinions of the State Council on Strengthening Key Works of Environmental Protection stipulated to perfect the insurance system of environmental pollution liability, and carry out the compulsory liability insurance pilot of environmental pollution.

Green trade: the export rebates have been canceled for some products with "high pollution or high environmental risk", and their processing trade has also been banned.

The environmental protection check for listed companies and industry: since 2007, the MEP has issued some files, such as Circular on Further Regulating the Environmental Protection Check for the Production and Operation Companies of the Heavy Pollution Industry to Apply for Listing or Refinancing, Guidance on Strengthening Supervision and Management of the Listed Company of Environmental Protection, Circular on Further Strictly Regulating Environmental Protection Check of the Listed Company and Strengthening the Supervision after the Environmental Protection Check and Circular on Further Regulating Supervision and Management of the Listed Company and Strictly Implementing Environmental Protection Check, which explicitly stipulated that the listed enterprises of 14 industries engaged in thermal power and steels etc. shall carry out the environmental protection check, the listed companies shall be urged to rectify through supervision after the environmental protection check, and continuously improved their environmental behavior; the listed companies shall be required to disclose environmental information to the public, so as to promote public participation and social supervision; the capital shall be guided to invest in the resources-saving and environment-friendly industries and enterprises through the environmental protection check before listing, so as to promote the transformation of economic development mode. Both Circular of General Office of the State Council Forwarding Guidance of the MEP and other Departments on Strengthening Prevention and Control of Heavy Metal Pollution and Opinions of the State Council on Strengthening Key Works of Environmental Protection explicitly stipulated to perfect the environmental protection check system for the company which are listed, refinanced or recapitalised in the first time, strictly conduct the environmental protection check for the listed enterprises.

Since 2001, the MEP has continuously carried out environmental protection checks for citric acid, monosodium glutamate, leather, rare earth, iron and steel, starch (starch sugar), alcohol industries, etc. The industrial environmental protection check system has been linked up with the environmental protection preferential policies, and interacts with the related departments; based on the industrial environmental protection check, the enterprise in compliance with environmental protection requirements can be supported with the priority during handling some administrative permits, such as the EIA approval, the import and export examination of solid waste and hazardous chemicals, and environmental protection funds supports, etc..

In July 2008, during the video conference of the national rural environmental protection convened by the State Council for the first time, Vice Premier, Li Keqiang put forward some policy measures for "promoting governance by means of awards, replacing subsidies with awards". The Central Finance Department set up special funds for rural environmental protection for the first time, totaling 4 billion CNY has been granted within three years so that above 6 600 villages and towns are supported to carry out the comprehensive environmental improvement and demonstrative ecological construction, above 24 million rural population directly benefited from it.

NO.6

Recommendations

Implement more ambitious air emission reduction targets capable of achieving ambient quality objectives already adopted; manage a wider range of air pollutants, including VOCs and toxic substances.

Implementation

In order to further improve atmospheric environment quality, the Chinese Government amended some discharge standards in view of the pollution characteristics and environmental capacity, etc. For example, new Ambient Air Quality Standard was issued in 2012, which adjusted classification of the ambient air function zone in this revision, revised the pollutants and their limit values.

The new Standard became compatible with the international practices, adding some monitoring indexes, such as the concentration limit of fine particles ($PM_{2.5}$), and 8 hours average concentration limit of ozone (O_3). In 2011, the MEP revised and issued Emission Standard of Air Pollutants For Thermal Power Plants, which adjusted the concentration limits for atmospheric pollutants discharged by the thermal power plant.

The atmospheric pollution control: the Chinese Government has established an integrated mechanism for regional pollution control, established an evaluation system for the regional air environment quality in the key integrated atmospheric pollution-control areas, and carries out co-ordination control for various pollutants. In 2010, the General Office of the State Council issued Guidance on Promoting the Integrated Atmospheric Pollution Control and Improving the Regional Air Quality; the atmospheric pollution control mechanism shall be further improved during the 12th Five-Year Plan period which focuses on Beijing-Tianjin-Hebei, Yangtze River, etc. so as to implement the co-ordinated control of multiply pollutants.

The total emission reduction of atmosphere pollutants: it increased pollutants category for total amount control, expanded the key fields for emission reduction, and strengthened the measures for emission reduction. The Program of the 11th Five-Year Plan for National Economy and Social Development of the Chinese Government stipulated SO_2 emission reduction target of 10%, and the requirements for the supporting target responsibility system and assessment mechanism. With efforts over five-years, China's 11th Five-Year Plan emission reduction target has been completely realised, total SO_2 emissions in 2010 was 21.851 million tonnes, fell by 14.29% in comparison with that in 2005, and over-fulfilled 10% of reduction commitment. The Program of the 12th Five-Year Plan for National Economy and Social Development of the Chinese Government further stipulated the total emission control targets, energy consumption and CO_2 emissions per GDP by the end of 2015 are reduced by 16% and 17%, SO_2 8% and NO_x 10% respectively.

The range of pollution control is being constantly expanded, the Program put forward to strengthen the volatile organic pollutants and toxic gas control during the 12th Five-Year Plan period. The 12th Five-Year Plan for Heavy Metal Pollution Comprehensive Control was approved in 2011 which specified that emission of heavy metals such as Pd, Hg, Cr and Ca, as well as metalloid as in key areas by the end of 2015 shall be reduced by 15% in comparison with that in 2007. Beijing, Shanghai, and Guangzhou etc. are trying to incorporate VOCs into the conventional monitoring scope. Premier Wen Jiabao pointed out in the 2012 Government Work Report that some monitoring items, such as $PM_{2.5}$ shall be conducted in some key areas, such as Beijing-Tianjin-Hebei, Yangtze River Delta and Pearl River Delta, municipalities directly under the central government and provincial capitals by the end of 2012, and promoted in all cities over prefecture level by the end of 2015. In 2010, the MEP with other departments jointly issued the Guidance on Strengthening of the Dioxin Pollution Control, which provided that "the reduction and control measures shall be fully implemented in the key industries, such as iron ore sintering, arc furnace steelmaking, regeneration of non-ferrous metals, and waste incineration, a comparatively perfect dioxins pollution control system and long-term supervision mechanism shall be established by 2015, dioxin emission intensity of the key industries shall be decreased by 10%, so as to the growth trend of dioxin emissions shall be basically controlled". In order to ensure the control target could be realised, the MEP organised some special investigations, such as, special investigation on the Ozone-Depleting Substance (ODS), special investigation on persistent organic pollutants, and special investigation on mercury pollution emission sources. In recent years, the MEP together with many departments conducted special enforcement inspection for some toxic substances, such as heavy metals, chemicals environmental management and hazardous wastes.

Strengthen the pollution control of motor vehicles: NO_x emission reduction of motor vehicles has been jointly promoted through comprehensive measures, including optimising the urban traffic, strictly implementing the emission standards of motor vehicles, promoting the environmental protection standards of motor vehicle, accelerating to phase out "yellow-label vehicle", and improving quality of fuel oil. The Reply on the Implementation Date of the 4th Stage of Limits of National Motor

Vehicle Emission Standard specified that spark ignition engine and vehicles which do not meet China-IV standard requirements cannot be sold and registered since 1 January 2011. In recent years, China-IV standard has been applied in Shanghai, Beijing, Guangzhou, Shenzhen and Nanjing in advance.

NO.7

Recommendations

Further improve the quality of monitoring data needed for effective air quality management and widen their scope (*e.g.* sources and pollutants).

Implementation

In order to improve quality of monitoring data, the new revised Ambient Air Quality Standard stipulated the stricter rules on the sampling method and effectiveness of data statistics, and updated the standards of monitoring analysis method. For monitoring capacity building, the MEP added Department of Monitoring in 2008. By the end of the 11th Five-Year Plan, the national environmental protection system has established 2 587 environmental monitoring stations, formed a 4-level environmental monitoring network of the state, province, city and district; the national air quality monitoring network consists of 338 cities over prefecture level from the original 113 key environmental protection cities; total 1436 national air quality monitoring sites have been set all over the country to monitor the six major pollutants with ozone, carbon moNO$_x$ide and PM$_{2.5}$ increased.

The MEP issued the Three-Year Action Plan for Environmental Monitoring Quality Management (2009-2011) in May 2009, which specified to actively promote construction of the environmental monitoring system, and implement the Three-Year Action Plan. In 2010, "the First Session of the National Competition of Environmental Monitoring Professionals and Technicians" was held. In 2011, "the National Practice Activity of Environmental Emergency Monitoring" was held.

In order to ensure the quality of environmental monitoring data, the MEP carried out the automatic monitoring capacity building for key pollution sources. Up to now, 349 pollution sources monitoring centres have been build at 3 levels of ministry, province and city, which can monitor 15 559 key pollution sources. In addition, the MEP has formulated some regulations, such as Measures for Automatic Monitoring and Management of Pollution Sources, Measures for Auditing Validation of

Automatic Monitoring Data, Measures for Managing Operation of Pollution Source Automatic Monitoring Facilities, and Measures for Site Supervision and Inspection of Pollution Source Automatic Monitoring Facilities, so as to improve validity of the environmental management data. At present, there are 6 063 industrial SO_2 discharge outlets, 4 329 industrial NO_2 discharge outlets with the automatic monitoring facilities, which can effectively monitoring pollution sources of the industrial waste gas. Automatic monitoring device must directly network with the competent departments of environmental protection for real-time data transmission. For pollution sources without the automatic monitoring device installed or automatic monitoring device not networked with the competent departments of environmental protection, the competent departments regularly carried out the manual monitoring, the monitoring frequency of national key enterprises is no less than once every quarter.

Through strengthening four aspects of capacity building, including the automatic monitoring for national key monitoring enterprise, supervisory monitoring on the pollution sources, environment supervision and enforcement, environmental information and statistics, the MEP are trying to establish a set of scientific system for transmission, verification and analysis on total amount data of major pollutant, a set of environment monitoring system combined the supervisory monitoring of pollution source and on-line automatic monitoring on the key pollution sources, as well as a set of strict and feasible evaluation system for total emission reduction.

NO.8

Recommendations

Develop and implement a national transportation strategy that recognises the environmental externalities of transport and takes an integrated approach to private and public transport; streamline the institutional framework for developing sustainable transport systems; use a mix of regulation and economic instruments (*e.g.* taxes) to give citizens incentives for rational transport decisions.

Implementation

In February 2008, the National People's Congress amended and promulgated Water Pollution Prevention and Control Law of the People's Republic of China, which firstly defined that information on water environmental status shall be released by the departments of environmental protection; In July of the same year, "Three Provisions" of the State Council further straighten out division of responsibilities among the departments of environmental protection, water conservancy and so on. Specifically, the department of environmental protection exercises the unified supervision and administration of water pollution control; the departments of water administrative at all levels are responsible for allocation and utilisation of water resources; the department of urban construction is responsible for construction and management of sewage treatment facilities; and the department of health is responsible for monitoring drinking water hygiene and disease epidemic prevention.

The Chinese government has implemented Standards for Drinking Water Quality (GB 5749—2006) since July 2007, which increased water quality indicators from the original 35 items to 106 items, added 71 items and revised 8 items. The MEP also values monitoring water quality of drinking water sources. In 2011, Premier Wen Jiabao pointed out in the government work report that: during the 11th Five-Year Plan period, construction of the rural agriculture infrastructure shall be sped up, so as to complete dam reinforcement for 7 356 large and medium-sized reservoirs and key small reservoirs to eliminate risks, solve issues of drinking water safety for 215 million in the rural population. Concerning health care, capacity building of the primary medical and health services shall be strengthened. The National finance shall allocate special funds to rebuild or build 23 000 township health clinics, 1 500 county hospitals, 500 county traditional Chinese medicine hospital and 1 000 county maternity and child care hospitals, as well as set up 2 400 community health service centres.

Continuously strengthen investment and management of the urban water supply and sewage treatment facilities: In 2007, the MoF issued Interim Measures for Capital Management by Means of Replacing Subsidy with Awards for the Supporting Pipe Network of the Urban Sewage Treatment Facilities, and decided the Central Finance to set up a special reward subsidy fund to support the central and western regions for incorporate their supporting pipe network of urban sewage treatment facilities into the 11th Five-Year Plan national planning, so as to enhance the urban sewage treatment capacity. In 2008, Article 44 of the new revised Water Pollution Prevention and Control Law of the People's Republic of China explicitly stipulated that: "the operating entity of urban sewage treatment facilities shall provide paid services to the polluters for sewage treatment according to the state regulations, and charge for sewage treatment, so as to ensure normal operation of the centralised sewage treatment facilities". During the 11th Five-Year Plan period, above 2 000 new sewage treatment plants increased totally with newly increased sewage treatment capacity more than 65 million tonnes/day; By the end of 2010, China has built 2 832 urban sewage treatment plants totally with the total treatment capacity of 125 million tonnes/day, which is doubled in comparison with that of the 10th Five-Year Plan period. The capacity of public water supply facilities of the planned cities and county cities in whole country increased 33 mn cm^3/day; the length of pipe network increased 222 100 km; and population of water utilisation increased by 96 million during the 11th Five-Year Plan period. By the end of 2010, the urban population (the planned cities, county cities and towns) with water supply facilities reaches 630 million with a coverage rate of 90.3%.

Strengthen supervision of the urban sewage treatment plant operation: The former Ministry of Construction issued the Opinions on Strengthening Supervision of the Urban Sewage Treatment Plant Operation in 2004; the MOHURD (Ministry of Housing and Urban-Rural Development) have also issued an industry standard of Technical Specification on Operation, Maintenance and Safety of Urban Sewage Treatment Plant in 2011. In addition, the MEP releases a list of all urban sewage treatment plants and informs the major pollutants discharged beyond the standard for the social public supervision every year. The sewage treatment plants are regulated by year according to the plan of energy conservation and emission reduction. During the 11th Five-Year Plan period, the urban sewage treatment rate increased from 52% to 77%. And meanwhile, the MEP incorporated all urban sewage treatment plants into the list of national key enterprises to be monitored, and required all of them to install on-line monitoring system and network with the department of environmental protection; the MEP also issued Circular on Strengthening Audit and Account of Emission Reduction From the Urban Sewage Treatment Plants (HB [2008] No. 90); and together with the relevant departments promulgated Policy on Sludge Disposal and Pollution Control Technology from the Urban Sewage Treatment Plants (CJ [2009] No. 23) to put forward the specific requirements for strengthening supervision of the urban sewage treatment plant operation, construction of automatic monitoring system and central control system, and sludge disposal. Since 2007, the MEP implemented the regional limited approval to some areas of which urban sewage treatment plant are constructed lag, the policy of sewage charge hasn't been implemented in place, the actual sewage treated amount is less than 60% of design capacity after the sewage treatment plant has been built for one year, or the built sewage treatment facilities haven't been operated for no reason.

NO.9

Recommendations

Continue efforts to improve water pollution control and efficiency in water use by industry; increase the rate of pollution charges and abstraction charges; ensure that treatment plants are efficiently managed; link abstraction and discharge permits to total load planning, while maintaining minimum flows and river quality objectives.

Implementation

Water Pollution Prevention and Control Law of the People's Republic provides that: "the local people's government shall reasonably plan industrial layout, require the enterprises which cause water pollution to upgrade their technology; adopt comprehensive control measures; increase rate of water recycling; and reduce waste water and pollutant discharge. The outdated technology and equipment which seriously pollute water environment shall be phased out; it is prohibited to newly construct production project which isn't in conformity with the state industrial policy or seriously pollutes water environment." Since 2008, MEP has successively issued a series of Cleaner Production Standard for wastewater discharge from key industries, such as chemical fiber, wine making, oil refining, coal selecting, tanning and electroplating; In order to strengthen management of lead accumulator and secondary lead industry, Circular on Strengthening Pollution Control for the Lead Accumulator and Secondary Lead Industry was issued in 2011 to require the new-build project related to lead must have a define data source of total lead pollutants emission.

Industrial water fees are being continuously improved, the related policies has been further improved and implemented, such as different water price and sewage treatment fee. In order to improve water use efficiency, Department of Water Conservancy has prepared the industry water quota, and implemented total water amount control at water shortage areas. Apart from this, the Chinese Government has implemented the stepped water pricing system in each industry, and improved the criteria for water use. The stepped water pricing system is also implemented in the living sector, linking total water amount with water price. The Polluter-Pay Principle will be fully implemented; the charge system of sewage treatment shall be continuously improved, the charge system shall gradually meet the requirements of stable operation of sewage treatment facilities.

The 12th Five-Year Plan for National Environmental Protection proposed that key areas and industry shall strengthen water pollutant emission reduction. Total emission control for nitrogen and phosphorus shall be implemented in the eutrophic lakes and reservoirs, and some coastal areas of the East China Sea and Bohai Sea where red tides occurs easily. The COD and total ammonia and nitrogen emission controls shall be promoted for the papermaking, the printing and dyeing, and the chemical industry with percentage cut down for no less than 10% in comparison with that in 2010. New projects only for expanding capacity shall be strictly controlled for papermaking, printing and dyeing, tanning, pesticides, and nitrogenous fertiliser, etc. in the Yangtze River Delta and Pearl River Delta. It is prohibited to newly construct non-ferrous metals, papermaking, printing and dyeing, chemical industry, tanning projects in some key river sources.

Since the 11th Five-Year Plan period, China has established the target responsibility system for total emission reduction and the corresponding performance evaluation mechanism; as an important tool to realise the total emission reduction target, the pollution discharge permit system has played a positive role in the implementation of total amount control planning and the annual plan. During auditing and accounting total emission reduction of major pollutants, the water balance has calculated for water intake, water use and sewage discharge of the enterprises. Water Pollution Prevention and Control Law of the People's Republic of China revised in 2008 stipulates that China will implement the total amount control system and pollution discharge permit system for key water pollutants. Regulations on the Pollution Discharge Permit which is being drafted by the MEP combines pollution discharge permit system with total amount control system closely. The permit will define the criteria for concentration of the discharged pollutants and indexes for total amount control, the permit holder whose total discharge amount is more than emissions criteria or indexes shall be punished according to the law. In order to further strengthen the pollution discharge permit system, the 12th Five-Year Plan for National Environmental Protection explicitly stipulated that "the pollution discharge permit system shall be fully implemented".

NO.10

Recommendations

Continue efforts to improve water pollution prevention and water efficiency in agriculture, and to establish water user associations responsible for recovering the cost of providing irrigation water; improve monitoring and collection of groundwater abstraction charges; take measures to halt overexploitation of groundwater aquifers; prevent agricultural pollution run-off into aquifers, rivers and lakes (*e.g.* buffer zones along rivers and lakes, treatment of intensive livestock effluents, efficient application of agro-chemicals); phase out fertiliser subsidies.

Implementation

In recent years, in order to strengthen the agricultural pollution control, the Chinese Government carried out the first National Census of Pollution Sources in 2007, preliminary disclosed the agricultural pollution status. Circular of General Office of the State Council on Forwarding Opinions of the State Environmental Protection Administration and Other Departments on Strengthening the Rural Environmental Protection put forward the specific requirements for the national agriculture and rural environmental protection. For the agricultural non-point source pollution control, the Central Government has invested 14.2 billion CNY to conduct the standardised transformation for scale farms (community) of pig and cow, including excrement disposal facilities. At present, Regulation on livestock and poultry pollution control is being drafting; the soil testing and formulated fertilisation technologies have been promoted in area over 1.1 billion *mu*, there are 545 counties (farms) in China which have implemented the project of soil organic matter ascension subsidy, strictly restrain to use pesticide with high toxicity and high residual; 33 kinds of pesticides have been successively phased out, the professional systematic control of pests and diseases and the area of green prevention has reached 650 million *mu*.

In 2008, the Chinese Government set up the special fund rural environmental protection to encourage the comprehensive improvement of the rural environment and ecological construction through "promoting treatment by reward and replace subsidies with reward". From 2008 to 2010, the special fund totaled 4 billion CNY, which led local input of nearly 8 billion CNY, supporting more than 6 600 villages and towns to carry out the comprehensive environment improvement and construction of ecological demonstration villages and towns, above 2 400 rural population directly benefited from it.

The 12th Five-Year Plan put forward higher requirements for agricultural and rural environmental protection. Agriculture, especially for livestock and poultry breeding has been brought into the scope of emission reduction, and meanwhile, the MEP and the MoA (Ministry of Agriculture) have carried out some co-operation on the livestock and poultry pollution prevention and emission reduction of agricultural source. In order to deeply promote the national groundwater pollution prevention, the MEP with other relevant departments jointly prepared and issued Planning for National Groundwater Pollution Control. The special stamp shall be set for the Environmental Impact Assessment on groundwater, the specific requirements was put forward for selecting sites of large-scale livestock and poultry farms, and the forbidden areas are delimited for the breeding. At the same time, the specific requirements have been also put forward for safety discharge after excrement from livestock and poultry breeding is treated. The Ministry of Land and Resources together with relevant departments formulated Planning for National Land Subsidence Control (2011-2020), this Planning was issued and implemented in March 2012, which defined objectives and tasks of land subsidence control in China in the future 10 years with overlift groundwater control taken as the major measures.

NO.11

Recommendations

Strengthen and further develop an integrated river basin management approach to improve water resources and water quality management, and to provide environment-related services more efficiently (*e.g.* flood and drought prevention, soil and water conservation, biodiversity protection, support for recreation and tourism); give greater weight to the protection of aquatic ecosystems (*e.g.* renaturation of rivers and lakes banks, protection of wetlands); foster stakeholder participation (*e.g.* representatives of economic sectors, environmental NGOs, experts, administration).

Implementation

In 2002, the revised Water Law of the People's Republic China established the long-term planning system for water supply and demand; strengthening implementation and supervision of the comprehensive basins planning. Water Pollution Prevention and Control Law of the People's Republic of China was take into effect since 1 June 2008; it defines the environmental responsibility of the local government, which provides a basic guarantee to strengthen water pollution prevention and control in China.

In view of the increasingly scarce water resources and water ecological deterioration in recent years, the Chinese Government put forward a plan to rehabilitate rivers, lakes and seas, implementing the strategy of comprehensive management for water resources and water environment, strengthening co-ordination and division of tasks among the sectors. In addition, in view of the regional characteristics of water resources distribution in China, some great projects of water conservancy have been implemented, such as the South-To-North Water Diversion Project, so as to balance the space allocation of water resources. And meanwhile, the protection plans have been specifically formulated for large rivers and large water sources to strictly restrain all kinds of development activities and provide eco-compensation to the subjects who suffered losses due to water ecological protection.

In order to ensure implementation of various policies and measures, the Chinese Government has established a strict mechanism for target decomposition and examination. Especially in the field of environmental protection, the Central Government established the appraisal system for special planning implementation of water pollution control in some key basins, so as to evaluate the local government on trans-provincial section water quality and special planning implementation with the related assessment results released to the public; the responsibilities of local government are further clarified, which effectively promoted the basin water pollution control. In addition, the agencies for regional environmental supervision have given full play in strengthening overall co-ordination of regional watershed environmental problems; The National Joint Conference of Departments of Environmental Protection was adjusted and improved, the Joint Conference of Departments for Key Basin Water Pollution Control was established to strengthen communication and co-operation between the department and the local government.

In order to improve the level of water pollution control, China carried out "major science and technology projects of water pollution control and management" with 32 projects and 230 subjects launched during the 11th Five-Year Plan period; 3.2 billion CNY of the central fiscal funds has been fully allocated in place, and a batch of problem have been broken through which provides the technical supports for water pollution control.

In 2008, the MEP released the Planning for Water Pollution Control in Key Basins, including the Huai River, Hai River, Liao River, Chao Lake, Dianchi Lake, the Upstream and Mid-Stream of Yellow River (2006-2010), namely the 11th Five-Year Special Planning for Water Pollution Control with the objective that the centralised drinking water sources of six key basins, including Huai River, Hai River, Liao River, Chao Lake, Dianchi Lake, the upstream and mid-stream of Yellow River, can receive the necessary management and protection; the trans-provincial section water quality can be improved obviously; key industrial enterprises can fully and stably achieve the discharge to satisfy the standards; the urban sewage treatment can be significantly improved, total amount of water pollutant can be effectively controlled; the capacity can be significantly enhanced for basin water environment regulation, early warning for water pollution and emergency response. From 16 March-3 April 2011, after the MEP with relevant departments under the State Council jointly assessed the 11th Five-Year Special Plan for water pollution control, the water pollution control of in key basins obtained remarkable achievements with compliance rate of water quality, completion rate of planning project and investment improved significantly.

During the 12th Five-Year Plan period, when the basin pollution control will continue to highlight the key points, the coverage shall be expanded to all large rivers, lakes and seas; the pollution control shall be carried out by region; the pollution in key units shall be controlled with priority. Through some measures, such as the fiscal and tax preference and preferential project, some places shall be encouraged firstly to change the situations of serious basin water pollution so as to rehabilitate it. In 2010, in order to strengthen management of Liaohe Reserve, Liaoning Provincial Government organised to form Administration of Liaohe Reserve. Based on the 11th Five-Year Plan, the MEP together with the relevant departments under the State Council jointly issued the Planning for Water Pollution Control in Key Basin (2011-2015) in May 2012 with planning scope includes 10 river basins of Songhua River, Huai River, Hai River, Liao River, the upstream of Yellow River, Tai Lake, Chao Lake, Dianchi Lake, Three Gorges reservoir area and its upstream, Danjiangkou Reservoir area and its upstream, the goal is to realise the stable water quality of the urban centralised surface drinking water source and achieve the functional requirements by 2015; water environmental quality can be improved obviously for the trans-provincial section, the seriously polluted urban water body and tributary; the eutrophication of lakes can be reduced; the compliance rate of water function area can be further improved; The ecosystem of Dianchi Lake can be obviously improved; Liao River Basin takes the lead in recovery from pollution control to the ecological restoration; The total amount of major water pollutants discharge and total amount of pollutants into the river can be constantly lowered; as well as capacities for water environment monitoring, early warning and emergency response can be improved significantly.

The Law of the People's Republic of China on Water and Soil Conservation taken into effect in 2011 specified the policies and measures for prevention and control of water and soil erosion as: "The measures, such as engineering measures, plantation and protective measures, shall be taken in the water erosion area according to the actual situations with small basins formed by natural gully and hilly land on both sides as a unit, so as to conduct comprehensive control of water and soil erosion on the slope farmland and gully."

China attaches great importance to wetland conservation. Planning for Implementation of National Wetland Protection Projects (2005-2010) was issued and implemented during the 11th Five-Year Plan period. Presently, the wetland protection network system has been preliminary established on the basis of wetland nature reserve and wetland park; and the central financial special subsidy for wetland protection is also established to subsidise the important international wetland, wetland nature reserve and the national wetland park, as well as wetland monitoring and ecological recovery. In the recent Five-Years, the Chinese Government has input more than 3 billion CNY to restore wetland for nearly 80 000 hectares. In addition, the Chinese Government put the Ramsar Convention in place, strengthening the role of the National Committee in performing Ramsar Convention in China, smoothly implementing some international co-operation projects between China and Australia, Germany, USA, etc..

Regarding water environment management, the Chinese Government encourages the public participation from all circles of the society, including experts from NGOs and departments. For example, in order to enhance the science of the planning and improve implementation effect of the planning, public participation was emphasised during the formulation of the planning for water pollution prevention in key river basins organised by the MEP with some means adopted, including network voting, consultation meetings and planning information disclosure, which promoted participation of the stakeholders. In addition, the government actively fed back supervisory suggestions put forward by NGOs during implementation of the planning.

NO.12

Recommendations

Further encourage sustainable water use through: i) institutional integration of water quality control and of water investments (*e.g.* at national and other relevant levels of government); ii) market-based integration with further progress in the transition towards full cost pricing of water services, while giving attention to the special needs of the poor and of the West; iii) clarifying and securing the rights to extract, allocate and use water, in the context of water legislation and land tenure reform.

Implementation

China is a country short of water resources and the government attaches great importance to water resources management. In 2011, the Chinese Government promulgated the Decision of Central Committee of the Communist Party of China and the State Council on Speeding up Reform and Development of Water Conservancy (ZHF [2011] No. 1) which proposes to strengthen the support of capacity of water conservancy and realise sustainable utilisation of water resources.

In order to promote sustainable water use, the Chinese Government has adopted a series of measures in recent years. On 21 February 2006, the State Council issued Administrative Regulations on Collection of Water Intake Permit and Water Resources Fee, which is of great significance in strengthening the centralised management and supervision of the national water resources, promoting conservation, protection and reasonable exploitation and utilisation of water sources, as well as constructing a resource-conserving society. On 9 April 2008, the MWR (Ministry of Water Resources) promulgated Administrative Measures for Water Intake Permit to strengthen management of water intake permit and standardise application, approval, supervision and administration of water intake. In January 2009, the MWR, the MoF and the NDRC (State Development And Reform Commission) jointly issued Measures for Use and Management of Water Resources Fee Collection, 31 provinces, autonomous regions and municipalities directly under the central government all began collecting water resources fee and strengthened management of the collection. In 2011, the MoF and the MWR jointly promulgated the Interim Measures of the Central Government for Use and Management of Water Resources Fee which improves the efficiency of water resources fee utilised by the Central Government. In addition, in July 2011, the MoF and the MWR jointly issued the Circular on Relevant Matters about Construction Funds for Farmland Water Conservancy Accrued from Land-Transferring Income, which requires all local governments charge 10% of the land-transferring income as construction funds according to the specified criteria since July 1st 2011, which is specially used for construction of farmland water conservancy facilities, mainly supporting construction of farmland water conservancy in major grain-producing areas, the central and western areas, old revolutionary base areas, ethnic minority areas, border areas and poverty-stricken areas.

The Program of the 12th Five-Year Plan for National Economy and Social

Development of the Chinese Government provided to implement the most stringent management system of water resources. The State Council issued Opinions on Implementing the Most Strict Management System of Water Resources (GF [2012] No. 3) in 2012 to put forward to establish the bottom line for development and utilisation of water resources and water use efficiency, as well as limiting water function containing pollution, and strictly implementing total water amount control.

In order to protect the rights of water bodies and deeply promote the water rights system and water pricing reform, the Chinese Government has carried out reform pilots of water rights and water pricing system at all levels across the country, strengthened the paid use and conservation of water resources through implementing step water pricing system, and compensated the damaged subjects through implementing the fiscal transfer payment and eco-compensation at the underdeveloped areas, especially the upstream of large rivers. Some provinces have issued special papers to standardise eco-compensation; for example, Guangdong Province formally promulgated Methods of Eco-Compensation in Guangdong Province in 2012, which divided eco-compensation into the basic compensation and incentive compensation; among of them, the former is to ensure expenditure needs of the basic public service; while the incentive compensation is linked with effects of key ecological function areas protection and the ecological environment improvement, the better ecological protection is, the more rewards obtained.

NO.13

Recommendations

Foster the move towards a circular economy by focusing on waste reduction, reuse of waste material and waste recycling, and related targets; require provincial and local governments to adopt and implement comprehensive waste management plans (including accurate verification of volumes of waste – municipal, industrial and hazardous – generated and treated) covering elements of the waste hierarchy.

Implementation

Since the 11th Five-Year Plan, China gradually strengthened policy formulation for waste treatment and resources reuse, as well as implementation of the related policy. In March 2007, the MoF, NDRC, Ministry of Public Security, original Ministry of Construction, State Administration for Industry and Commerce, and the former

SEPA have jointly issued the Measures for Recovery and Management of Renewable Resources. In September 2007, the former SEPA issued Administrative Measures for Electronic Waste Pollution Control to strengthen management of electronic wastes. The Ministry of Industry and Information Technology (MIIT) has recently issued the first batch of the Directory of the Advanced Applicable Technologies for Comprehensive Utilisation of The Renewable Resources, covering comprehensive utilisation industries, including waste electrical and electronic products, waste tire rubber, waste metal and waste glass, waste plastics and waste textiles, building and agriculture and forestry waste, waste paper and other six categories of products. Circular Economy Promotion Law of the People's Republic of China was effective since January 1st 2009, which promotes recovery of the renewable resources and regulates development of renewable resources recovery industry. In addition, the MoC has implemented the pilots of renewable resources recovery system since 2006, there are total three batches of 90 urban pilots carried up to now.

During the 12th Five-Year Plan period, China has set higher management objectives for solid waste. The Program of the 12th Five-Year Plan for National Economy and Social Development provides that: "comprehensive utilisation rate of industrial solid waste shall reach 72%, and resource output rate shall increase by 15%; the resources recycling recovery system shall be improved." In order to implement the Circular of the State Council on Printing and Distributing Work Scheme of the 12th Five-Year Plan for Energy Conservation and Emission Reduction, the MOHURD issued Opinions on Accelerating the Development of National Green Building in April 2012, which specified the energy and resources consumption level during construction and use of the buildings in China in 2020 shall reach the level of developed countries at the present stage.

Meanwhile, each department conducted division of labor and co-operation, arranged and deployed resources recycle. In April 2012, the General Office of the State Council issued Circular on Responsibility-Assigning Scheme among Key Sectors during Establishing a Complete Advanced Recycling System for Waste Commodity, which requests the MoC jointly with the relevant departments to ensure that all key tasks and policy measures can be implemented in place. In May 2012, in order to promote the sustainable, healthy and harmonious development of the industry of renewable resources recovery, MoC issued Circular on Conducting Investigation on the Renewable Resources Recovery Industry and decided to carry out investigation on situations of the renewable resources recovery industry.

For comprehensive solid waste management, MEP, MoC, NDRC, the General Administration of Customs and AQSIQ (General Administration of Quality Supervision) jointly issued the Administrative Measures for Solid Waste Import in 2010; the MEP printed and issued Administrative Provisions on Environmental Protection for Imported Solid Wastes as Raw Materials, which required to strictly conduct a technology review and environmental management for importing solid wastes available as raw materials; give full play to the comprehensive functions of environmental protection, so as to promote transformation of economic development mode; collaboratively promote industrial restructure and pollution reduction; and regulate and enhance the pollution control level of renewable resource industry. In 2009, the MEP issued Standard for Comprehensive Ecological Industrial Park (Trial), which specified the park must be provided with the collect system of waste water, solid waste (including waste electronic products and so on) and central wastewater treatment facilities. In 2010, the NDRC and MoF jointly issued Circular on Construction of the Urban Mineral Demonstration Base, planning to built about 30 "urban mineral" demonstration bases in the whole nation through Five-Years efforts. The key is to promote the recycling, scale utilisation and high-efficiency of such key "urban mineral" resources as scrap electrical and mechanical equipment, wire and cable, electrical appliances, automobiles, mobile phones, lead-acid battery, plastics and rubber.

NO.14

Recommendations

Accelerate the pace of extending waste treatment capacity by building treatment infrastructure and establishing systems for the collection, reuse and recycling of waste (*e.g.* separate collection of household waste), including in rural areas.

Implementation

Since 2006, China has issued a series of documents to promote recycling work. At present, the MoC has successively carried out the pilot of renewable resources recovery system in 90 city all over the country; On August 20th 2008, the State Council issued the Administrative Regulations on Recycling and Disposal of Waste Electrical Appliances and Electronics Products in order to regulate recycling and disposal of waste electrical appliances and electronics products; In 2009, the MoC and MoF further put forward to accelerate the establishment of renewable resources recovery system. The MHURD organised classification and recovery of domestic garbage at the large- and medium-sized cities. In April 2012, in order to implement Opinions of the General Office of the State Council on Establishing a

Complete Advanced Recycling System for Waste Commodity, ACFSMC (All China Federation of Supply and Marketing Co-operatives) issued Opinions on Accelerating Construction of Waste Commodity Recycling System for the Supply and Marketing Co-operatives to promote construction of waste commodity recycling system for the supply and marketing co-operatives, which put forward an objective that the total amount of waste commodity recycling in whole system accounted over 60% of that in whole society by end of the 12th Five-Year Plan.

In recent years, China has carried out specific waste recovery in some specialised fields. For example, in June 2009, the General Office of the State Council issued Circular on Forwarding Implementation Plan of NRDC and other Departments on Promoting the Expansion of Domestic Demand and Encouraging Trading in Car and Appliance, and carried out the trade-in pilots in some provinces and cities, such as Beijing, Tianjin, Shanghai, Jiangsu, Zhejiang, Shandong, Guangdong, Fuzhou and Changsha. The MoF has formulated the Measures for Primage of the Trade-in Home Appliance, which plays a positive role in promoting the effective utilisation of the resources. In 2011, the MEP issued Circular on Carrying out a Special Inspection on Business Entities of Waste Lead-Acid Battery (HBH [2011] No. 470) to strengthen the environmental management of the entities of which are engaged in utilisation and disposal of waste lead-acid battery with operation permits of hazardous wastes to prevent heavy metal pollution.

In view of the domestic garbage recycling, the former Ministry of Construction issued Administrative Measures for Urban Domestic Garbage in 2007, which put forward the specific requirements for cleaning up, collection, transportation and treatment of domestic garbage. In April 2012, the General Office of the State Council issued the 12th Five-Year Plan for Construction of Harmless Disposal Facilities for National Urban Domestic Garbage, which provided that by 2015, the domestic garbage in the municipalities directly under the central government; the provincial capitals and cities specifically designated in the State Plan shall realise harmless treatment; the domestic garbage processing rate of the planned cities shall reach above 90%; all counties shall have their own garbage harmless treatment capacity; the harmless treatment rate of domestic garbage in the county town shall reach more than 70%, the newly increased capacity of harmless treatment facilities for urban domestic garbage shall reach 580 000 tonnes/day.

For the rural domestic garbage treatment, the MoA has organised the rural areas in whole country to construct the rural clean engineering demonstration, and implement collection, disposal and recycling of wastes from the agricultural production and farmers living since 2007, the recycle and utilisation rate of crop straw, people and livestock feces, domestic garbage and sewage in the demonstration villages reached up to 90%.

NO.15

Recommendations

Formulate enforcement plans for different sectors (*e.g.* households, large- industry, small and medium-sized enterprises) and types of waste.

Implementation

Presently, the Chinese Government implemented a mandatory plan for hazardous waste, medical waste, radioactive waste and dangerous chemicals.

For the hazardous waste and medical waste, the Chinese Government has formulated Law of The People's Republic of China on Prevention of Environ Mental Pollution Caused by Solid Waste and Administrative Measures for the Permit of Hazardous Wastes Operation to implement the permit system of dangerous wastes operation for the entities which engaged in the collection, storage, utilisation, and disposal of hazardous waste. The entities which discharge the hazardous waste must prepare the management plan for hazardous waste in accordance with the relevant national regulations. In 2011, the MEP and MoH (Ministry of Health) jointly issued Opinions on Further Strengthening Regulatory of the Dangerous Waste and Medical Waste to propose more strict regulatory requirements for hazardous waste and medical waste. In 2008, the MEP launched modification of the Administrative Measures for Sheet for Transferring the Hazardous Waste to continuously carry out the special inspection on the entities of hazardous waste incineration and strengthen management of the entities which generate wastes. In September 2007, the former SEPA issued Administrative Measures for Pollution Control of Electronic Waste to prevent environment pollution of electronic waste and strengthen management of electronic waste. In order to ensure that dangerous waste can be properly disposed, there are about 1500 entities in China with permits of dangerous wastes operation which are engaged in collection, storage, utilisation, disposal of hazardous waste by the end of 2011.

For the radioactive waste management, the Chinese Government has implemented Regulations on Management of Radioactive Waste since 1994, which defined the management goal and basic requirements for design and operation of each link, including generation, collection, treatment, transportation, storage and disposal of the radioactive waste. In view of safety management of the radioactive waste, the Chinese Government issued Regulations on Safety Management of Radioactive Waste (Decree no. 612 of the State Council) in December 2011, which put forward the requirements for treatment, storage and disposal, supervision and management of radioactive waste.

For the hazardous chemical management, the Chinese Government has implemented a registration system of the hazardous chemicals. Administrative Measures for Registration of the Hazardous Chemicals was revised in 2012 which specified the registration contents, registration procedures, and the relevant responsibilities of registered enterprises, as well as requirements for supervision and management. At present, the MEP is studying to formulate Measures for Registration of Environmental Management of the Hazardous Chemicals in order to reduce the release of dangerous chemicals into the environment and prevent environmental risk.

NO.16

Recommendations

Streamline the allocation of responsibility for the management of different types of waste; ensure that waste facilities operate efficiently and comply with standards; further develop workable regulations and policy instruments for waste management; improve the collection of waste data and develop tools to evaluate the effectiveness of waste management policies at national and provincial levels.

Implementation

For waste management, according to Environmental Protection Law of the People's Republic of China and Law of The People's Republic of China on Prevention of Environ Mental Pollution Caused by Solid Waste amended and promulgated in 2004, the departments of environmental protection are responsible for supervision and management of solid waste and hazardous waste; Governments and Departments of Housing and Urban-Rural Development at all levels are responsible for construction and management of domestic garbage, hazardous waste and medical waste disposal facilities; MoC and ACFSMC are responsible for the recycling of renewable resources.

In order to further strengthen management of the solid waste and radioactive waste, the Chinese Government has successively issued Administrative Measures for

Environmental Pollution of Electronic Waste, National Directory of Hazardous Waste (2008 Revision), Administrative Registrations on Recycling and Management of Waste Electrical Appliances Electronics Products, and Registrations on Safety Management of the Radioactive Waste in recent years. For supervision and administration of the environmental pollution control of solid waste, Law of The People's Republic of China on Prevention of Environ Mental Pollution Caused by Solid Waste provided that: "the construction project shall build the necessary supporting facilities for environmental pollution control of solid waste, which must be designed, constructed and put into operation together with the main part of the project. In addition, the competent department of environmental protection and other departments of regulating the environmental pollution control of solid waste shall have the rights to conduct on-the-spot inspection on the entities related to the environmental pollution control of solid waste within their scope of jurisdiction according to their respective duties.

By the end of 2008, all provinces in China had established the solid waste management centres and hazardous waste disposal centres, some cities also established the solid waste management centres; such as Shenzhen; All cities have established the medical waste disposal centre, and correspondingly issued the charging measures for medical waste.

Concerning the management of waste data collection, the Chinese Government has established information-release mechanism for waste in the large- and medium-sized cities, which regularly releases information related to solid waste to the public every year. In 2009, the MEP began construction of management information system for the national solid waste, which will further improve the capacity of waste data collection after the system is established. In order to effectively evaluate waste management effect, the MEP conducted the nationwide standardised assessment on hazardous waste since 2010, and promoted each local government to implement the related policies, laws and regulations on the hazardous waste management.

NO.17

Recommendations

Establish financing mechanisms with a mix of public and private financing, and move to charging for waste services more progressively in less developed areas; improve the collection rate of waste charges and set them at a level consistent with the government's aim to achieve a circular economy.

Implementation

For hazardous waste, in order to implement the Law of The People's Republic of China on Prevention of Environmental Pollution Caused by Solid Waste and Administrative Registrations on Medical Wastes in place, strengthen the hazardous waste disposal and promote the industrialisation of hazardous waste disposal, the NDRC, the former SEPA, MoH, MoF and the Ministry of Construction jointly issued Circular on Implementing the Charges System for Hazardous Waste Disposal to Promote the Industrialisation of Hazardous Waste Disposal in 2003 to require all local government reasonably establish the charge criteria for dangerous waste disposal, the charge criteria shall be verified in accordance with the principle that the cost for hazardous waste disposal can be compensated and a reasonable profit be made. The cost for hazardous waste disposal mainly includes transportation fees, material fees and expenditure on power, maintenance, and depreciation of facilities equipment, etc. occurred during collection, transportation, and storage treatment of dangerous waste. Guangdong, Shandong and other provinces have all implemented the hazardous waste charge system. Presently, the market mechanism of waste disposal has been gradually established, which has already formed a diversified investment mode combining the State investment with the local and private investment.

Regarding electronic waste, in order to support Administrative Registrations on Recycling and Disposal of Waste Electrical Appliances and Electronics Products implemented since 2011, Administrative Measures for Collection and Use of Fund for Waste Electrical Appliances and Electronics Products Disposal was released 2012, which provided that: "the charge shall be collected according to the amount of the electrical and electronic product which the producer has sold, or the amount which the consignee or its agent who are engaged in importing the electrical and electronic product has imported, at the same time the subsidies shall be provided to the relevant enterprises in accordance with the amount of the waste electrical and electronic products which actually disassembled and disposed. The subsidy criteria shall be: 85 CNY/set for TV set; 80 CNY/set for refrigerators; 35 CNY/set for washing machine; 35 CNY/set for room air conditioner; and 85 CNY/set for microcomputer".

Concerning domestic garbage, the Chinese Government has actively promoted treatment and charge for domestic garbage. In recent years, China urban domestic garbage collection and transportation network has improved; the quantity and capacity of domestic garbage disposal facilities are also rapidly improved. By the end of 2010, the annually cleaned amount of urban domestic garbage in national planned cities and counties reached 221 million tonnes, the harmless treatment rate of domestic waste reached 63.5%; among them was 77.9% for planned cities. In 2007, the Ministry of Construction issued Administrative Measures for Urban Domestic Garbage which specified: "if the entity and individual did not pay the urban domestic garbage disposal fee according to the regulation, the entity can be fined certain amount less than three times of the payable urban domestic garbage disposal fee and not more than 30 000 CNY fine, the individual can be fined certain amount less than three times of the payable urban domestic garbage disposal fee and not more than 1 000 CNY fine". The Beijing released Administrative Regulations for Beijing Urban Domestic Garbage in 2012, which stipulated the economic penalties to the entity which does not conduct garbage classification, treatment, recovery as required.

On the whole, upholding the principle of "reducing, reuse and being harmless", a series of policies and measures have been issued in recent years to gradually improve waste recycling and treatment. In the central and western underdeveloped areas, government departments have strengthen guidance on investment into construction of waste treatment facilities. For waste resource recycling, the Chinese Government has issued some preferential policies to encourage the private capital enters into the resources recycling field. At present, some resources recycling enterprise which are invested and operated by the private capital have emerged in China.

NO.18

Recommendations

Provide the informal sector (freelancers) with equipment, organisational assistance and training to continue collection and recycling under improved hygienic and environmental conditions, as part of waste management plans.

Implementation

At present, the Chinese Government has carried out extensive training for farmers' waste recycling. For example, China Resources Recycling Association has set up specialised agencies responsible for training farmers' waste recycling, which has carried out a series of training activities, such as "serve to new countryside and training farmer recycling workers", widely promoting the local industry association of renewable resources recycling training for above 80 000 farmers workers, the supply and marketing co-operatives at all levels, and the relevant units directly under the ACFSMC also actively carried out the relevant training; through the professional training, practitioners engaged in waste recovery have mastered the knowledge on distinguishing waste, improved the skills for separation and disassembly, and changed the pattern of disordered recovery and secondary pollution.

ACFSMC puts forward an objective that above 80% of the urban community establish a standard recovery site by the end of the 12th Five-Year Plan; innovate process for recycling and operation, extensively carry out the flexible and diverse recovery techniques, such as telephone appointment, online recovery, and replacing the old with the new; promote the recycling enterprises establish a stable relationship with sectors and units, production and processing enterprises for renewable resources recovery, and establish a diversified and multi-channel convenient recycling system, so as to further smooth recovery channel and promote the network operation level.

At the same time, NGOs, such as Friends of Natural and Global Village, also carried out various activities to improve the capacity of the urban and rural residents to collect waste.

NO.19

Recommendations

Raise awareness of waste management and efficient resource use among the public, small and medium-sized enterprises, and industry.

Implementation

In recent years, the MEP and its related departments have carried out various forms of advertising activities, so as to raise the waste management consciousness of the public and related units in view of solid waste management; in 2010, a training for Administrative Measures for Collection and Use of Fund for Waste Electrical Appliances and Electronics Products Disposal and related policies promotional session was held; in view of the policies and regulations on solid waste management, such as newly issued Administrative Measures for Solid Waste Import and Administrative Provisions on Environmental Protection for Imported Solid Wastes as Raw Materials, training sessions were held to improve the environment-friendly policy and legal consciousness of the enterprises engaged in waste import, and to promote the effective utilisation of renewable resources.

In 2011, the MoC decided to carry out pilots of the demonstration project for upgrading the enterprise engaged in the scraped car recovery and disassembly in 10 cities, and provided the related demonstration projects the capital support which may reached 50% of total investment of the project at most, so as to enhance the comprehensive utilisation of resources and environmental protection of the industry engaged in the scraped car recovery and disassembly. On 1 June 2012, the Electronic Waste Recycling and Disposal Branch of China Resources Recycling Association together with hundreds of enterprises launched an initiative to call on standardising recovery, environment-friendly disassembly, law-abiding business, and create a green homeland.

In order to strengthen the waste management consciousness of enterprises, Nokia and its partners have carried out the "Green-Box" program in China since 2005, above 700 service outlets of about 300 cities in China mainland have been set up the recycling bins for waste mobile phone and accessories. By the end of 2010, they have recovered above 160 tonnes of waste mobile phones and accessories, which recycle the renewable raw materials, such as metal and plastic through environment-friendly treatment by a professional company entrusted by Nokia.

In order to promote a resource-conserving and environment-friendly society, the government agency, the NDRC organises "National Publicity Week Activities" every year, which intensively promotes the national conditions of energy and resources, strengthening the resources risk and conservation consciousness of the public, and promote whole society further embodying energy saving, low carbon and green concepts into action.

NO.20

Recommendations

Modernise and implement legislation on nature protection, in particular adopt a law on the protection and management of Nature Reserves, notably favoring an increase of marine protected areas and of protected areas with higher protection status; consider ratification of the Bonn Convention.

Implementation

The Chinese Government is accelerating the ecological protection legislation. Nature Reserve Law of the People's Republic of China, Soil Pollution Prevention and Control Law of the People's Republic of China, Transgenic Biological Safety Law of the People's Republic of China, and Ecological Protection Law of the People's Republic of China are under formulation. The nature reserve legislation has been listed in the legislation program of the National People's Congress (NPC); the NPC Environment and Resource Panel is drafting Natural Heritage Protection Law of the People's Republic of China.

The MEP seriously implements the comprehensive management functions of nature reserve specified in the Regulations on Nature Reserves, and actively promoted construction and development of the national nature reserves. The MEP together with the relevant departments conducted evaluations on the national nature reserve management. Since launched in 2008, 231 national nature reserves of 22 provinces, autonomous regions and municipalities directly under the central government have been evaluated. By the end of 2011, there are 2 640 national nature reserves established in the whole country, a total area of nature reserves is about 149 km^2, accounting for about 14.9% of the total land area.

In December 2010, the General Office of the State Council issued a Circular on Implementing Relevant Management of the Nature Reserve, which required scientific planning development of nature reserves, strengthened management of nature reserve range and adjustment of function zones, strictly restrained development and construction activities related to the nature reserves, strengthened management of development and construction projects involved in the nature reserves, regulated management of land and maritime within the natural protection, and intensified supervision and inspection of the nature reserve. Besides, the Circular also stressed to strengthen the leadership and co-ordination in the future.

Construction of the nature reserve is realising transformation from number-scale type to the quality-benefit type. Besides, the MEP is launching implementation of remote sensing survey and evaluation on the national ecological environment variation within ten years (2000-2010), and issued Regionalisation of National Ecological Function, which further enhanced protection level of the regional ecological function.

NO.21

Recommendations

Enhance the capacity of national, provincial, prefecture and county level agencies to manage biodiversity protection of existing reserves and integrate nature conservation within economic and social development projects outside protected areas.

Implementation

China is a signatory to the Convention on Biodiversity, and has already sign the Convention on June 11th 1992. With agreement by the Environmental Protection Committee of the State Council, the former SEPA and the related departments jointly issued China Action Plan of Biodiversity Conservation in June 1994. At present, the seven major goals determined by the Action Plan have already realised basically, most of 26 priority actions have been completed. Subsequently, the Chinese Government has successively issued Outline of Development Plan for China Nature Reserve (1996-2010), Plan for Construction of National Ecological Environment, Outline of Plan for Protection and Utilisation of Biological Species Resources (2006-2020), and Outline of Action for China Aquatic Biological Resources. The competent departments of the related industry also issued and implemented a series of planning on nature reserves, wetland, aquatic biology, and livestock and poultry genetic resources protection respectively.

The MEP and above 20 departments and units jointly formulated China Biodiversity Conservation Strategy and Action Plan (2011-2030), this Plan was issued and implemented in September 2010, which defined China overall objectives, strategic mission and priority action of biodiversity conservation in future 20 years. In June 2011, the State Council approved the establishment of China Biodiversity Conservation Committee with Vice Premier Li Keqiang served as Chairman, the responsible leaders of 25 member units as members. The Committee established a long-term working mechanism for co-ordination of national biodiversity protection. The MEP also actively promoted "mainstreaming" of biodiversity protection, strengthened promotion, and promoted each departments and the local government incorporate the biodiversity protection into the relevant planning and action plan of the departments and the local government.

In addition, the Chinese Government has implemented "China Action of 2010 International Year of Biodiversity" and "China Action of the United Nations Decade on Biodiversity" with remarkable achievements obtained. The Chinese Government has also carried out about 230 large-scale promotional activities, influencing about 900 million people in 2010.

Natural protection has become an important factor that the Chinese Government will consider during the implementation of the economic and social development project beyond the protection zones. The Technical Guideline for Regional Environmental Impact Assessment on Development Zone provided that: when the environmental impact assessment shall be conducted on the development zone, the impact on ecological environment shall be analysed for the planning and implementation of development zone.

NO.22

Recommendations

Increase the financial and human resources for nature and biodiversity protection and further involve local residents in inspecting, monitoring and habitat enhancement, in the context of poverty eradication; diversify the sources of financing of nature conservation.

Implementation

Over the years, the governments at all levels and relevant departments have taken active measures to incorporate the construction and management of nature reserve into the plans for national economic and social development, and increased investment in the nature reserves, in accordance with the provisions of Regulations on Nature Reserve. In 1998, the MoF set up the national special fund to support all nature reserves to strengthen capacity building; the current amount of the fund reaches CNY 100 million every year. In 2001, the NDRC approved to implement "construction project of national wildlife protection and nature reserves", and conducted the overall planning and engineering construction arrangements for the national wildlife and nature reserve construction in the future 50 years.

During the 11th Five-Year Plan period, the State Forestry Administration fully implemented the forestry development strategy with priority to the ecological construction, accelerated transformation from the traditional forestry to modern forestry, focused on building the forestry ecology and industry system, and implemented the engineering driving and transformed the pattern of growth, so as to fully improve quality and efficiency of the forestry development, and constantly develop a various functions of forestry. The MEP has increased about 46% of the administrative formation since 2007; the staff working for natural and biodiversity protection have been supplemented correspondingly.

China has also taken measures to encourage local residents to participate in management of the nature reserves. When the nature reserve is established, the opinions of community residents shall be solicited as possible, if it is involved in the contracted land of the residents, the entrusted management agreement shall be signed with the contractor. In addition, many nature reserves have set up community condominium mechanism, the community residents are increasingly participating in the relevant protection and management.

NO.23

Recommendations

Develop the use of economic instruments related to nature and biodiversity protection, not only as income supporting measures, but also to reward the provision of environmental services.

Implementation

In China, the primary economic tools for development and utilisation of natural ecology and biodiversity is based on the eco-compensation policies. In August 2007, the former SEPA issued. "Guidance on Implementing Eco-Compensation Pilot" which provided guidelines for national eco-compensation. Presently, the eco-compensation tools in China mainly include investment in some important ecological engineering project, subsidies for returning farmland to forestry or returning grazing land to grassland, compensation to the ecological public welfare forest, national bonds to subsidise construction of the rural biogas facilities, subsidy policy for water conservancy of small farmland and water and soil conservation, etc. The area of forest received the Central Financial eco-compensation has reached to 1.259 billion *mu*; the grassland ecological protection covered by the subsidy and reward systems accounts for 64% of the natural grassland in whole country; all provinces (autonomous region, municipality directly under the central government) have established and implemented their margin system for mine-site environmental recovery. The implementation of the eco-compensation policy has played an important role in safeguarding the ecological security in China and promoting the equalisation of ecological environment services.

The Chinese Government is actively promoting eco-compensation pilot and legislation, and promulgated "Measures for Transfer Payment to the Key Ecological Function Zones". The Chinese Government began to implement the fiscal transfer payment system at the key ecological reserves, such as Three-River Source and Dong River Source since 2008. In 2010, China approved some coastal cities, such as Weihai in Shandong Province, Lianyungang in Jiangsu Province to be marine eco-compensation pilot areas. In 2010, the Chinese Government launched the wetland eco-compensation pilot and conducted eco-compensation for protecting more than 90 important international wetlands and wetlands nature reserves.

NO.24

Recommendations

Integrate long-term plans for rehabilitating and maintaining species and protected areas (including managing alien species) with land-use and river basin management plans, and any subordinate provincial, prefecture and country plans.

Implementation

Restoring and maintaining species and reserves has become an important aspect to be considered by the Chinese Government as well as the relevant provincial, city and county planning during the preparation of planning for land use and basin management, In 2004, the General Office of the State Council issued "Circular on Strengthening Protection and Management of the Biological Species Resources", which explicitly required that departments at the various levels make the planning for protection and utilisation of biological species resources. According to the requirement of the General Office of the State Council, the former SEPA actively organised the relevant ministries and commissions to study and formulate the planning for protection and utilisation of national biological species resources, and formally issued Outline for Protection, Utilisation and Planning of the National Biological Species Resources in 2007, which defined the overall objective and stage goals of the Outline for protection, utilisation and planning of the national biological species resources by 2020, and determined the key fields in the next 15 years, recent priority action areas and the priority projects for protection and utilisation.

All provinces, autonomous regions, municipalities directly under the central government also established their own biodiversity conservation co-ordination mechanism, set up committees of experts to prepare the provincial strategy and action plan for biodiversity protection. Some local governments carried out the protection and restoration of biodiversity and poverty alleviation demonstration, which combined biodiversity protection with poverty alleviation. Some areas and basins also prepared the local action plan of biodiversity protection in combination with national strategy for biodiversity conservation, such as Action Plan of Northwest Yunnan for Biodiversity Protection.

In addition, in order to strengthen, uniformly organise and co-ordinate protection and management of the biological species resources in China, the former SEPA organised to establish the Inter-Ministry Joint Conference on the Biological Species Resources Protection consisted of 17 relevant ministries and commissions under the State Council in 2003. Since Inter-Ministry Joint Conference was founded, the MEP has actively organised and co-ordinated various ministries and commissions in protection and management of biological species resources; and organised six sessions of inter-ministry joint conferences on the biological species resources protection by April 2012, which has laid a foundation for China's planning for protection and utilisation of biological species resources. In 2004, the MoA formed the National Invasive Alien Species Control Co-operation Panel consisting of seven relevant ministries and commissions of the State Council, the MoA formed Office of Alien Species Management and Research Centre of Invasive Alien Species Control, which laid the foundation for incorporating alien species management into the related planning and guiding to control the invasive alien species in China.

NO.25

Recommendations

Integrate the economic and social values of protecting habitats and species (*e.g.* ecological services, tourism development) within development of decision-making, in particular as parts of EIAs.

Implementation

The State Council promulgated "Regulations on the Planning for Environment Impact Assessment" in 2009, which defined to take "the overall effect of the planning implementation on the related regions, basins and sea ecosystem" as the content to be analysed, forecasted and evaluated in the planning EIA. The MEP takes the ecological protection as one of the important elements in making the related planning EIA and project EIA technical guidance. When the EIA is examined implementation approval for major development and construction project, the habitats and species protection is an important inspecting element.

In order to protect the habitats and species resources, the MEP together with other departments issued "Circular on Strengthening Adjustment and Management of Nature Reserves" in 2008. In 2010, the General Office of the State Council issued "Circular on the Relevant Works About Nature Reserve Management", which strengthened management of the nature reserve range, function zone adjustment and development, and construction projects of nature reserves. The MEP organised and formulated the Specification on Preparing Application Materials for National Nature Reserve Scope and Function Zone Adjustment, which strictly controlled the adjustment of national nature reserve, strictly implemented the ecological access of the construction projects involving national nature reserve, and further regulated review on the nature reserve adjustment. Each local government shall strictly manage the approval process for development and construction project of the nature reserve, and formulate and promulgate the administrative measures for development and construction projects of the local nature reserve and approval provisions for environmental impact assessment.

NO.26

Recommendations

Consider establishing an inter-ministerial group to examine how environment related taxes might be reformed to help achieve environmental policy objectives better.

Implementation

Depending on China's national conditions and characteristics of administrative management system, the environmental protection authorities have strengthened their co-ordination and co-operation with the related functional departments to establish a trans-government for environmental protection co-operation mechanism. At present, the Ministry of Environmental Protection, the Ministry of Finance and the State Administration of Taxation are jointly studying the environmental taxation and have developed a preliminary tax-for-fee scheme. The general principle of the scheme is to clear expense and profiting tax step-by-step. In accordance with the Polluters Pay Principle, the taxes for the pollution discharging enterprises will be launched when conditions permit. The central government has issued a policy to grant business income tax preference to the enterprises that purchase and actually use energy-saving and emission reductions equipment and environment protection equipment; improved the preferential policy of Value Added Tax for comprehensively utilised products that meet environmental protection requirements; and granted the preferential policy of value added tax to FGD byproducts and power generation from medical waste and sludge incineration. It levies consumption taxes on the consumer goods that consume resources considerably and pollute the environment; and considers environmental protection factors in vehicle and vessel taxes and grants preference to purchase tax of light cars.

In addition to the above taxes relating to environmental protection, China encourages the export of clean products and suppresses export of heavy pollution products by drawback and tariff policy adjustment in trade links to promote domestic resource saving and environmental protection. Since 2007, the Ministry of Environmental Protection has published the comprehensive environmental protection catalogue of 514 types of "high pollution and high environmental risk" products and heavy pollution processes, 42 types of environmentally friendly processes and 15 types of key environmental protection equipment for pollution emission reduction. Hence, the Ministry of Finance, the State Administration of Taxation and the General Administration of Customs have cancelled approximately 200 tax registration numbers of obsolete products and prohibit their processing trade.

NO.27

Recommendations

Increase and diversify the sources of environmental finance by fuller implementation of the polluter pays and user pays principles, and increase the effectiveness and efficiency of allocating public environmental expenditure.

Implementation

China is developing multi financing channels for environmental protection investment. The total investment in environmental protection in the 11th Five-Year Plan period is 2 160 billion CNY, 156.4 billion CNY of which is central financial funds, accounting for 7.24% total investment.

China is already implementing the Polluter Pay principle. The Regulations on the Management of Collection and Use of Waste Discharge Fees (Decree No.369 of the State Council) promulgated in 2003 provides that the polluters who directly discharge pollutants to the environment should pay waste discharge fees. In order to ensure the legal, complete and sufficient collection of waste discharge fees, the Method for Inspecting the Collection Work of Waste Discharge Fees (Order No.42 of the former State Environmental Protection Administration) was put into implementation in December 2007.

China now is actively advancing the reform of waste discharge fees and wastewater treatment fees mainly for the purpose of promoting the collection standards for the Polluter Pays and User Pays, suppressing and reducing the externality of environmental behaviors and facilitating standardised collection.

NO.28

Recommendations

Strengthen the institutional mechanisms for better integrating environment into economic and sectoral policies, and if possible, establish a Leading Group on environment or on sustainable development; fully implement the provisions in the EIAs law for assessing the potential environmental impacts of sectoral programmes.

Implementation

China has recongnised the important role of integrated environmental and economic decision making in protecting resource environment and optimising economic growth and all major ministries and commissions including the National Development and Reform Commission, Ministry of Environmental Protection, Ministry of Land and Resources, Ministry of Water Resources and Ministry of Agriculture have established effective working mechanism to co-ordinate the contradiction between economic development and environmental protection. Since the 11th Five-Year Plan period, the environmental goal has become an important restrictive index for the national economic and social development plan. In the 11th Five-Year Plan period, total emission of COD and SO_2 have dropped 12.45% and 14.29% respectively, both successfully accomplished emission reduction tasks. Total emission reduction goal forces, in a reverse way, structural readjustment and elimination of backward production capacities, including closedown and suspension of 720 000 000 kW small thermal power units, elimination of 121 720 000 tonnes iron making production capacity, 69 690 000 tonnes steel making production capacity and 330 000 000 tonnes cement production capacity. The draft Environmental Protection law of the People's Republic of China under revision also definitely proposes the concept of "harmonious economic growth and social development and environmental protection".

In 2006, the concept of "Three Changes" (change in pattern of demand, change in industrial structure and change in element input) proposed in the Report to the Seventeenth National Congress of the Communist Party of China was well implemented in the 11th Five-Year Plan period and afterwards. In June 2007, the State Council established the National Leading Group to Address Climate Change and Energy Conservation and Pollutant Discharge Reduction led by Premier Wen Jiabao to study and deliberate the major policy issues and co-ordinate and solve major problems in work. In 2008, on the basis of the former Environment Protection Leading Group of the State Council, the inter-ministry joint conference system for environmental protection was established to co-ordinate the decision making mechanism for major environmental issues such as biological species and trans-regional watershed management. Local government also established a similar departmental co-ordination mechanism and Ningbo City in Zhejiang, for example, established the linkage mechanism for mutual co-ordination between environmental protection authorities and public security departments.

In the Seventh National Conference on Environmental Protection held in 2011, Vice Premier Li Keqiang sufficiently recognised the significant achievements in the environmental protection work in the 11th Five-Year Plan period, systematically analysed the prominent problems and deep-seated contradictions in the present environmental protection, firmly proposed to adhere to "protecting in development and developing in protection" to actively explore a low cost, high benefit, low emission and new sustainable environmental protection road, practically solve the prominent environmental problems which affect scientific development and harms the people's health and comprehensively push forward China's environmental protection.

At the same time, the provisions on environmental impact assessment in departmental development plans in China's environmental impact assessment laws and regulations have been fully implemented. In August 2009, the Regulations on Plan Environmental Impact Assessment was promulgated as Decree No.599 of the State Council and put into implementation on October 1st in the same year. Now implementing environmental impact assessment has become an important link for formulating a departmental development plan. Inner Mongolia, Chongqing, Shenzhen and Dalian etc. have enhanced the implementation of the environmental impact assessment plan through local legislation. Since 2011, the Ministry of Environmental Protection, jointly with the National Development and Reform Commission and the Ministry of Transport, issued a document to define the management requirements for planning environmental impact assessment in related fields. The total emission reduction goal has become a precondition for examination and approval of environmental impact assessment and one-vote veto is executed. Henan etc. has also issued the management methods for related pre-examination and approval.

NO.29

Recommendations

Continuously to establish national targets to achieve key environmental objectives, taking scientific, economic and social analysis into account.

Implementation

Though scientific economic and social analysis China has set a strict national goal to realise the harmonious development between environment and economy. Environmental issues drew more and more attention in the 11th Five-Year Plan period in all fields. The "Outline of the 11th Five-Year Plan for National Economic and Social Development of the People's Republic of China" proposed the restrict indices of 20% reduction of energy consumption per unit GDP and 10% reduction of total emission of major pollutants SO_2 and COD.

The 12th Five-Year Plan for National Economic and Social Development of the People's Republic of China proposes that by 2015, energy consumption per unit GDP and CO_2 emission will be reduced by 16% and 17% respectively, COD and SO_2 reduced by 8% and ammonia nitrogen and NO_x reduced by 10%. At the same time, the Chinese Government also starts regulating the pollutants closely related to public health such as heavy metal, VOC, $PM_{2.5}$ etc. in the key range of environmental management.

To ensure the operability of planning objective, the decomposition process of total emission reduction task undergoes "two ups and two downs". The Ministry of Environmental Protection works out a 12th Five-Year Plan emission reduction guide to all provinces, autonomous regions, municipalities directly under the Central Government as well as the 8 major central enterprises, and then the local regions report the first draft of the 12th Five-Year Plan emission reduction plan in their respective regions based on this guide, which is called the "up"; then the Ministry of Environmental Protection organises experts to examine the first draft and sends it to all regions, which is called the first "down"; the local regions revise the first draft accordingly, which is called the second "up"; and finally after approved by the Central Government, it is eventually determined and issused as the 12th Five-Year Plan emission reduction plan, which is called the second "down".

NO.30

Recommendations

Further improve health and living standards, particularly in less developed areas, by reducing the ratio of people without access to good environmental services (safe water, basic sanitation, electricity); taking account of affordability constraints, give higher priority to water infrastructure investment in development strategies (*e.g.* for the poorer central and western China).

Implementation

In 2006, the 11th Five-Year Plan for National Safeguarding Rural Drinking Water Project was put into implementation, and in 2006 and 2007, total investment of 29.1 billion CNY was actually completed in the country and drinking water safety problem for totally 65 220 000 people was solved. In 2008, the Ministry of Environmental Protection organised the National Basic Environmental Survey and Assessment Work for drinking water source areas to ascertain the basic environmental conditions of drinking water source areas in towns and typical villages and townships in China, to establish and improve the basic information of centralised drinking water source areas, and to scientifically evaluate the basic environmental conditions of drinking water source areas in the country.

In March 2012, the Central Government approved the 12th Five-Year Plan for National Safeguarding Rural Drinking Water Project which requires further accelerating construction while continuously consolidating the built project achievements and completely solving the drinking water safety problem for 298 000 000 rural people and 114 000 rural schools so that the population covered by rural centralised water supply in the country is increased to 80%.

Since the 11th Five-Year Plan period, China has been actively advancing the medical and health system reform and basic health care system building; strengthening medical service supervision and management and making efforts to solve the high cost and difficult medical treatment problem. Urban and rural health conditions have changed and people's health level been further improved. Within less than 2 years after the "Electricity Supply to Every Household" project was implemented, electricity use for 974 000 no-electricity households of 3 600 000 people has been solved. The new type rural co-operative medical service system tried in 2003 basically covered the whole country in 2010 and its implementation greatly reduced the peasants' medical treatment costs.

The Ministry of Finance gradually increased the balanced transfer payment of the state revenue to the central and west regions to shrink the gap of local standard financial revenue and expenditure in the central and west regions, and push forward the inter-regional basic public service equalisation. The special transfer payments such as energy saving and environmental protection, new energy resources, education, talents, medical treatment, social security and poverty alleviation through development are mainly inclined towards the central west region.

The National 12th Five-Year Plan for Environmental Protection lists the environmental infrastructure public service project as one of the major environmental protection projects, including construction of urban domestic pollution and hazardous waste treatment and disposal facilities, and urban drinking water source-area safety assurance projects.

In 2011, the central government issued No.1 Order to emphasise water conservancy construction and make efforts to ensure a breakthrough in water conservancy investment mechanism; annual average investment of the whole society in water conservancy in the next 10 years can be doubled compared with 2010. Implement the most stringent water resource management system and determine the three "Red Lines" of water resource development and utilisation control, water use efficiency control and pollution discharge restriction for water function zone. For deep level restriction on rapid and good water conservancy development, it is emphasised that a new breakthrough should be made in water conservancy system.

NO.31

Recommendations

Continue to improve environmental information by developing and using indicators of environmental performance, environment-related economic information and analysis, and environmental accounting tools such as Material Flows Accounts; expand the coverage of environmental information (*e.g.* to diffuse pollution, toxic substances, hazardous waste); continue to improve consumer protection and public access to environmental information.

Implementation

China's basic environmental management system emphasises the role of environmental performance management and restrictive indices for emission reduction. Among them, "creating a national environmental protection model city" and "quantitative overall urban environmental quality evaluation" as two major means have effectively improved urban environmental quality. Enhance environmental infrastructure construction and give full play to the provincial level environmental protection authorities. Disseminate the "creating a national environmental protection model city" and "quantitative overall urban environmental quality evaluation" systems to the whole country. Implementing and including these two systems in local governments' performance evaluation is conductive to further mobilise their enthusiasm and comprehensively advancing urban environmental protection work in our country.

In 2007, the State Council arranged and conducted the first National General Survey of Pollution Sources for understanding the environmental conditions in China and basic environmental information of various enterprises and institutions, establish and improve various major pollution source archives and pollution source information databases at all levels, accurately learn about the emission of pollutants and set a stage for correctively judging environmental situation, scientifically working out the environmental protection policies and plans, and practically improving environmental quality. According to environmental management need the 12th Five-Year Plan environmental statistical system is reformed in statistical framework, statistical range and statistical method on the basis of general survey of pollution sources to provide a basis for information releasing.

China is gradually establishing and utilising environmental performance index, environment-related economic information analysis and environmental accounting tool, which has enlarged environmental information coverage and made the public more aware about environmental information. Since 2004, the Ministry of Environmental Protection has been carrying out green national economic accounting work. Now it is conducting the study on the listed companies' environmental performance evaluation and their environmental information disclosure. In August 2008, the new edition of National Hazardous Waste Inventory was formally implemented; on May 1st 2008, the former State Environmental Protection Administration promulgated and implemented the Provisions of the People's Republic of China on the Disclosure of Government Information in China and the Measures for the Disclosure of Environmental Information to further standardise and promote the governments' and enterprises' environmental information disclosure.

The Ministry of Environmental Protection continuously issues "China Statistical Yearbook on Environment" and "Environmental Quality Bulletin" and publishes pollution sources monitoring information on the homepage of the Ministry of Environmental Protection website at the appropriate time. Environmental protection authorities at all levels also publish the information on environmental quality by means of quality bulletin and media. In addition, the Guidelines for Publishing the Information on Environmental Pollution Prevention and Control for Solid Wastes in Large-and Medium-Sized Cities provides that the competent administrative departments for environmental protection under the people's governments in large- and medium-sized cities should publish the information of the previous year on environmental pollution prevention and solid waste control in their own cities.

NO.32

Recommendations

Consolidate and strengthen information on health and the environment and develop a national health-environment action plan; implement the most cost-effective measures; promote pollution release and transfer reporting by enterprises; build capacity to report on exposures of specific population groups to environmental health risks (*e.g.* occupational health, health impacts near polluting facilities, children's health).

Implementation

Information on environment and health has been further improved. The Hazardous Waste Transfer Manifest system has been implemented for many years. To better fulfill the tenet for protecting the people's health, the Ministry of Health and the former State Environmental Protection Administration, jointly with the 16 ministries and commissions, formulated and issued the National Environment and Health Action Plan (2007-2015) in November 2007. This action plan was the first programatic document for driving China's environment and health work to develop in a scientific way. Since the action plan was issued, the Ministry of Environmental Protection and the Ministry of Health have sufficiently played the role of government organisation and leadership and actively promoted public participation. Regarding capacity building, they enhance management and technical supporting ability through many approaches; in scientific research, they increase fund investment and accelerate the speed at which the achievement in scientific research provides technical support to environmental management; in promoting local work, they push local environmental protection authorities to earnestly implement the action plan.

The first National General Survey of Pollution Sources started in 2007 which listed discharge of pollutants such as heavy metal and their environmental risk as the key survey items. In 2009, the Ministry of Environmental Protection issued the Notification on Strengthening the Environmental Management Registration Work for Import and Export of Toxic Chemicals to further regulate the examination and approval for environmental management registration for import and export of toxic chemicals, reinforce the supervision and management of imported and exported toxic chemicals flow, and comprehensively prevent and control the environmental risks of toxic chemicals in production, use, storage, transport and disposal. In 2010, the Ministry of Environmental Protection issued and implemented the Provision on the Environmental Administration of New Chemical Substances to effectively prevent the impact of new chemical substances on human health and ecological environment and eliminate the production of POPs substances. In addition, in accordance with the Basel Convention and the Law of the People's Republic of China on Prevention of Environmental Pollution Caused by Solid Waste, legal restraint is imposed on the solid waste that is trans-boundary moved and can be used as raw materials.

China now allows restricted imports of 10 major waste types such as waste paper, waste metal, waste hardware and electric appliance, waste ship and waste plastic and makes the environmental problems closely related to the people's health the key work points, *e.g.* PM$_{2.5}$, heavy metal and waste lead acid cell. It also carries out special prevention and control plan and rectification actions for the severely afflicted areas of electronic waste dismantling such as Guiyu in Guangdong.

NO.33

Recommendations

Further expand environmental education and raise envrionmental awareness, particularly among young people.

Implementation

To enhance environmental education in middle and elementary schools and promote green school transformation and upgrading, on the basis of national green school creation activity launched by the former State Environmental Protection Administration jointly with the Ministry of Publicity and Ministry of Education in 1996, environmentally friendly school pilot work was started in 2010 and environmental education base construction work was also carried out.

With the "World Environment Day on July 5th" as the core part, China carries out environmental education promotional activities for different topics every year, including exhibition, show, environmental documentary film production, annual personage selection for Green China and China Baogang Environmental Protection Prize Appraisal activities. On June 5th 2009, the Ministry of Environmental Protection, jointly with the Environment and Resources Protection Committee of the National People's Congress, Committee of Population, Resources and Environment of the CPPCC, Ministry of Science and technology, Ministry of Education, Central Committee of the Communist Youth League and All-China Women's Federation, initiated the "1 000 Young Environmental Friendship Heralds Action" activity to choose 1 000 "young environmental friendship heralds" in the country. These volunteers go deeply into communities, institutions, schools, enterprises, parks and plazas to propagandise and explain energy saving and emission reduction to the public and by one to thousand propagation, drive a million young people to actively participate in environmental protection. The representatives of the 1 000 Young Environmental Friendship Heralds were interviewed by Vice Premier Li Keqiang and they went to Durban to attend the United Nations Climate Change Conference in December 2011 and exhibit Chinese young people's thinking and action for addressing climate change to the international community.

On April 22nd 2011, the six departments of the Ministry of Environmental Protection, Ministry of Publicity, Civilization Office of the Central Communist Party Committee, Ministry of Education, Central Committee of the Communist Youth League and All-China Women's Federation jointly printed and issued the "National Action Programme for Environmental Publicity and Education (2011-2015)" which requires building the social action system participated by the entire people, including environmentally friendly schools, social practice base construction for environmental education in middle and primary schools.

NO.34

Recommendations

Continue enhance cooperation to work with NGOs and the public to achieve environmental policy goals; strengthen co-operation and partnerships with enterprises and commit social responsibility.

Implementation

The Ministry of Environmental Protection has always been attaching great importance to the promotion of non-governmental environmental protection organisations to the environmental protection cause and it actively supports and gives full play to NGO' role. In recent years, a large group of environmental protection NGOs that are influential both at home and abroad in China has emerged, *e.g.* Friends of Nature, Earth Village and Institute of Public and Environmental Affairs.

The Ministry of Environmental Protection uses various methods to strengthen exchange and communication with environmental protection NGO. For environmental protection NGO, the Ministry of Environmental Protection has determined the basic principles such as "active supporting and speeding up development" and "enhancing communication and deepening co-operation". In April 2012, the Ministry of Environmental Protection organised and held a work symposium for social environmental protection organisations and Vice Minister Pan Yue attended the symposium. More than 20 social environmental protection organisations including the Centre for Legal Assistance to Pollution Victims under

China University of Political Science and Law, Friends of Nature and Institute of Public and Environmental Affairs performed face to face exchange for how to promote the public and social environmental protection organisations to effectively participate in environmental protection. In June 2012, the Ministry of Environmental Protection organised environmental NGOs to attend the United Nations Conference on Sustainable Development held in Rio de Janeiro in Brazil, which pushed the international exchange and co-operation of environmental protection NGOs in China and exhibited the force and image of environmental protection NGOs in China. From central to local, the public-orientated interaction channel is established by means of media, Internet and communication etc..

The Ministry of Environmental Protection, in co-operation with the Legislative Work Committee of the National People's Congress, suggests that environmental public welfare lawsuit should be provided in the revised Civil Procedure Law. Now the draft has proposed that "for the behaviors that damage the social and public interests such as those that pollute the environment and infringe on broad consumers' lawful rights and interests, appropriate bodies and social groups can institute a law suit to the people's court".

In recent years, environmental protection NGOs has had a greater and greater influence in China's environmental protection cause. For example, Earth Village has developed simple social community sanitation maintenance and refuse classification to a present social group of some influence on the government and its main activities include: establishing green social communities and fostering ecological villages etc..

NO.35

Recommendations

Continue China's active engagement in international environmental co-operation, seeking to improve the effective and efficient use of: i) domestic resources; ii) international support mechanisms (*e.g.* the World Bank's Clean Development Fund, the Multilateral Fund under the Montreal Protocol, and the Global Environment Facility).

Implementation

In order to promote international environmental co-operation, the Chinese Government has established China Council for International Co-operation on

Environment and Development (CCICED) presently chaired by Vice Premier Li Keqiang. After establishment, the CCICED has successively organised many policy research projects involving many fields in environment and development and proposed several important policy suggestions. At the same time, the Foreign Economic Co-operation office established in 1989 has developed an operating pattern of mainly carrying out bilateral and multilateral co-operation in environmental protection projects and strengthening global environmental policy study as well as international consultation and service for environmental protection to further promote the China's foreign co-operation work for environmental protection.

In recent years, China has actively utilised international support mechanisms in combination with domestic resources to carry out international environmental co-operation in energy saving, greenhouse gas emission reduction and ozone depleting substances elimination and has made significant achievements.

Since July 2006, the Chinese Government has actively co-operated with GEF and obtained good global environmental benefit while promoting domestic environmental protection work. At the same time, the Chinese Government and private enterprises also provide some counterpart funding and physical goods and help projects generate good performances. The related Chinese Governments departments also co-operate closely and carefully implement GEF projects in the five fields including: biodiversity, climate change, persistent organic pollutants (POPs), international water areas and land deterioration in the light of the National 11th Five-Year Plan for environmental protection and national and economic development goals. They not only actively exert the "catalysis function" and "leverage" of GEF fund "seed money" but also promote the major environmental protection work such as energy saving, emission reduction and ecological construction. In the energy saving and emission reduction field, they are actively implementing the "energy saving air conditioner", "urban traffic development strategy co-operation and demonstration", "energy efficiency financing", "increasing thermal power plant energy efficiency" and "Beijing Green Olympics Electric Automobile" projects. After the "May 12th" Earthquake in 2008, the Ministry of Environmental Protection actively strived for about USD 2 000 000 GEF grants to implement the two emergency post-disaster reconstruction projects in biodiversity and persistent organic pollutants fields and obtained positive results.

The World Bank Clean Development Fund. The buyer of the clean development mechanism (CDM) project developed by the Ministry of Environmental Protection is clean development fund managed by World Bank. The project implementation and greenhouse gas emission reduction achievements are significant. By the end of

2011, the CDM project developed by the Ministry of Environmental Protection had obtained emission 137 000 000 t CO_2 equivalent certified by the United Nations, accounting for 29% total emission of all CDM projects in China certified in the same period and 17% certified total emission in the world. In addition to promote fund assignment, the co-operation projects with the World Bank also promote the introduction of HFC_{23} decomposition technology and while controlling and reducing greenhouse gases emission, they increase enterprises' environmental management level and achieve two win.

The Montreal Protocol multilateral fund. The Montreal Protocol multilateral fund is donated by developed countries and industrial transformation countries to mainly help developing countries fulfill ODS reduction obligations. Since 1991, the multilateral fund has cumulatively increased more than USD 2 billion and over 5 000 projects have been implemented in about 140 countries to support industrial transformation, technical assistance, information dissemination, training, and capacity building activities. Their priority in investment is the industrial transformation projects for direct ODS reduction. Up to now, China has cumulatively obtained 937 million USD Montreal Protocol multilateral funds and totally implemented more than 400 individual projects and 25 industrial plans to benefit over 3 000 enterprises. This plays a huge role in eliminating ozone depleting substances and protecting ozonosphere.

Sino-European biodiversity projects and watershed management projects, funded by EU are sufficiently utilised to promote domestic environmental protection work. The Sino-European biodiversity project has obtained 30 000 000 euro grants. It is the largest scaled biodiversity project sponsored by EU in oversea countries and also the international biodiversity co-operation project that has the largest fund scale, covers the largest territory and has the largest number of participating institutions and personnel in China. It was successfully completed in 2011. The project played an important role in promoting biodiversity protection and sustainable utilisation in China and became the successful Sino-European co-operation model in environmental protection field. The planned total investment for Sino-European watershed management project was 186 500 000 euro including 25 000 000 euro grants, 9 000 000 euro World Bank loans. The domestic implementing institution was the Ministry of Environmental Protection and Ministry of Water Resources. The project aimed to explore and implement the comprehensive watershed management mode consistent with the present global social, economic and natural changes, develop specific systems and procedures, carry out pilot work in the Yangtze River and the Yellow River watersheds and if possible, disseminate related experience to other regions on the basis of experience exchange with Europe and other regions in the world.

Other international co-operation projects: The Sino-Italian Co-operation Program for Environmental Protection Office was established for Sino-Italian co-operation. Totally, more than 90 co-operation projects have been implemented. It gave support to important activities and central and west development such as the Beijing Olympic Games, Shanghai World Expo, post-earthquake construction in Wenchuan, Sichuan and Yushu, Qinghai. At the same time, through lake pollution treatment, air monitoring and photovoltaic power generation projects, it drove Italian technologies and products into the Chinese environmental protection market, promoted inter-enterprise exchange and co-operation and related domestic technical level and realised mutual benefits and co-win. During Premier Wen Jiabao's visit to Italy in 2010, Sino-Italian environmental protection co-operation achievements were highly recognised by the two countries' premiers.

NO.36

Recommendations

Strengthen monitoring, inspection and enforcement capabilities in supportting the implementation of international commitments (*e.g.* on trade in endangered species, in forest products, in hazardous waste and in ozone-depleting substances, as well as on hazardous chemicals management, ocean dumping and fisheries management).

Implementation

Monitoring, supervision and law enforcement abilities are being developed to provide effective guarantee for fulfilling international commitments. The state revenue sets special funds for emission reduction of major pollutants to support environmental supervision and management ability building such as environmental monitoring, supervision, emergency response, information, promotion and education. In the 11th Five-Year Plan period, the cumulative investment in the country was 30 billion CNY including 12 billion CNY central investment. Meanwhile, the Ministry of Environmental Protection organised 6 regional environmental supervision agencies including North China, East China, South China, Northwest, Southwest and Northeast, added the Department of Environmental Monitoring and Satellite Environmental Application Centre and increased 90% staffs for China National

Environmental Monitoring Centre, which effectively enhanced national and local environmental supervision and management ability.

Concerning hazardous waste management, after the Basel Convention became effective in China on August 20th 1992, the Chinese Government has taken effective measures to strictly fulfill the Basel Convention. It has issued the National Hazardous Waste Inventory that complies with the Basel Convention, carefully executed the Law of the People's Republic of China on Prevention of Environmental Pollution Caused by Solid Waste and related laws and regulations, strictly examined and approved inspection formalities and issued special laws and regulations to strictly control trans-boundary movements of wastes. In 2008, the former State Environmental Protection Administration issued the Administrative Measures for Examination and Approval of the Export of Hazardous Wastes. Before the Rotterdam Convention on the Prior Informed Consent (PIC) Procedure for Certain Hazardous Chemicals and Pesticides in International Trade formally became effective to China on June 20th 2005, China had already included the chemicals and some pesticides involved in the Convention in the List of Banned or Strictly Restricted Toxic Chemicals and voluntarily executed prior information procedure for import and export of hazardous chemicals and pesticides.

China is an official party to the Stockholm Convention on Persistent Organic Pollutants and one of the first countries that signed the Convention in May 2001. Following accession to the Convention, China has carried out more than 30 international co-operation projects such as emission reduction and substitution demonstration of pesticide-like persistent organic pollutants, POP waste disposal and emission reduction technology demonstration for major dioxin-like pollution emission industries; issued over 30 management policies, standards and technical guidelines concerning POPs pollution prevention and control and convention fulfillment; realised complete elimination of 9 pesticide-like persistent organic pollutants including D.D.T.; disposed about 20 000 tonnes POPs waste and contaminated soil and safeguarded the people's health and environmental safety.

In order to implement the regulations that "Units discharging, collecting, storing, transporting, using or treating hazardous waste shall work out emergency plan and protection measures to be adopted in cases of an accident" in the Law of the People's Republic of China on Prevention of Environmental Pollution Caused by Solid Waste and guide the organising units for hazardous waste to work out the emergency plan and effectively respond to accidents, the former State Environmental Protection Administration worked out the Guide for Hazardous Waste Operating Units to Work out Emergency Plans in 2007.

Regarding bilateral and multilateral environmental monitoring co-operation, Russia and China jointly monitor trans-boundary water bodies including the Wusuli River, Khanka Lake, Argun River, Helongjiang River and Suifen River every year and China passes on water quality testing metres, gas chromatograph and active carbon to Russia without charge. In order to strengthen ocean dumping management, the Chinese Government has successively issued a series laws and regulations including Regulations of the People's Republic of China on Control Over Dumping of Wastes in the Ocean.

NO.37

Recommendations

Improve governmental oversight and environmental performance in the overseas Chinese corporations (in the spirit of the OECD guidelines for multinational enterprises).

Implementation

With the development of the global economy, environmental problems are drawing more and more attention. The Chinese Government attaches increasing importance to the environmental behaviors of its enterprises abroad. Related departments are improving the Guide for Environmental Behaviors of Chinese Enterprises Abroad to guide the enterprises to actively fulfill social responsibilities for environmental protection, safeguard good development of China's investment abroad and promote sustainable development in the regions invested in. Since 2007, the China Minmetals Corporation has issued China Minmetals Corporation Sustainable Development Report every year and evaluates the group's fulfillment with reference to the ten principles of the UN Global Compact.

Before this, the Chinese Government had seen the need for overseas services for environmental policies and taken some action. The State Forestry Administration, jointly with the Ministry of Commerce, issued A Guide on Sustainable Overseas Silviculture by Chinese Enterprises in August 2007, which initiated a new Chinese model of overseas forest harvesting.

In the 15th APEC Economic Leaders' Meeting in 2007, President Hu Jintao's suggestion to establish the Asia-Pacific Network for Sustainable Forest Management and Rehabilitation was written in the Sydney Declaration. The Ministry of

Commerce issued the Method for Management of Investment Abroad and Guide for Investment Countries in March 2009 to strengthen guidance service work for Chinese enterprises abroad and promote Chinese Enterprises to actively and reliably carry out investments abroad. In March 2003, the State Forestry Administration and the Ministry of Commerce issued A Guide on Sustainable Overseas Forests Management and Utilisation by Chinese Enterprises to actively guide and regulate Chinese enterprises' sustainable forestry activities abroad, promote the host countries' sustainable forestry development and safeguard Chinese Government's international image of a responsible big power.

Concerning foreign green credit, the China Banking Regulatory Commission issued the Green Guide Guideline in the beginning of 2012 to drive bank financial institutions to effectively prevent environmental and social risks and better serve economy through green credit. The Green Guide Guidelines provide that "for the overseas project to which credit is to be granted, we make a public commitment to adopt related international conventions or international standards to ensure that the operation of the project is kept essentially consistent with the international good practice."

NO.38

Recommendations

Develop partnerships with foreign enterprises to contribute to environmental progress through provision of training, technical support and cleaner technology; ensure that there in no sacrifice on environmental requirements for attracting foreign direct investments.

Implementation

In recent years, China has paid more and more attention to environmental protection, Chinese environmental protection laws and regulations and policy systems are developing and improving and environmental supervision and management ability has been enhanced greatly. China treats domestic and foreign enterprises equally without discrimination in pollutants emission and standards. China strictly executes various environmental management systems for foreign enterprises that invest in China and absolutely does not allow the attraction of direct foreign investment by reducing environmental admittance requirements. Under the supervision of social public opinions in the past years, China gives some punishments to the foreign-funded enterprises that have illegal environmental acts in China. For instance, the former State Environmental Protection Administration investigated and prosecuted Hitachi Construction Machinery (China) Co., Ltd. according to the law for its illegal and beyond-standard waste discharge behavior in 2007.

In 2009, the Ministry of Commerce and the Ministry of Environmental Protection jointly issued the Circular on Strengthening Energy-conservation and Environmental Protection Statistics of Foreign Investment, requiring that when transact the establishment and exchange of foreign-invested enterprises, all local competent commercial authorities should require enterprises to submit environmental impact assessment report to environmental protection authorities for examination and approval and they also add environmental protection indices to the foreign-invested enterprise examination and approval management system to increase foreign capital utilisation quality and comprehensively measure foreign investments' energy saving and environmental protection level.

NO.39

Recommendations

Intensify domestic and international co-operation to reduce transboundary air pollution in Northeast Asia by introducing cleaner coal technology, improving energy efficiency and switching fuel.

Implementation

To improve atmospheric environmental quality, for important atmospheric pollutants, the Chinese Government lists the major pollutants such as SO_2 etc. as mandatory emission reduction indices in the Outline of National Economic and Social Development Plan and adopts technical means including clean coal technology in actual facility reduction measures to increase energy utilisation efficiency and fuel conversion factor. After several years efforts, significant effect has been achieved. In the 11th Five-Year Plan period, total SO_2 emissions were reduced by 14.29%.

Over the last several years, through bilateral and multilateral co-operation, some international co-operation projects have been carried out to strengthen energy utilisation and capacity building and study pollution emission reduction policies, laws and regulations, *e.g.* Study on Policies for Promoting Circular Economy

Development in China technically assisted by the World Bank, Study on the Establishment of China Green National Accounting System etc. This is to enhance understanding at policy level and strategic height, advocate the concept of using clean energies, introduce and use advanced energy technologies and effectively increase energy utilisation.

The Ministry of Environmental Protection co-operated with the Asian Development Bank to conduct an SO_2 emission trading system study in 2008 in order to design SO_2 emission trading system with the help of advanced international concepts and enhance atmospheric pollution control management. It has actively carried out some energy saving and emission reduction pilot projects with bilateral countries, including the Sino-Italian Co-operation Ningxia Biomass Energy Power Generation Project, the Sino-Italian Environmental Protection Co-operation-Taiyuan Energy Efficiency Project and the Sino-Italian Environmental Protection Co-operation-China Sustainable Development Energy Efficiency Web Portal Project etc..

As a member country, China actively participates in various activities of the Northeast Asian Environmental Co-operation Mechanism (NEASPEC), including thermal power trans-boundary atmospheric pollution control, trans-boundary nature conservation projects etc. The Ministry of Environmental Protection takes the lead in implementing the "Establishing Northeast Asian Trans-boundary Region Nature Conservation Co-ordination Mechanism" project.

Now China's pollution emission reduction effects not only have improved its own environmental quality but also made contributions to regional environmental quality improvement.

NO.40

Recommendations

Ensure that the interim and final targets for the phase-out of ozone-depleting substances under the Montreal Protocol continue to be achieved on schedule.

Implementation

Since accession to the Montreal Protocol in 1991, China scrupulously abides by various provisions in the Protocol and through organising implementation management agencies to carry out convention policy study and multilateral fund project implementation activities. China has obtained significant progress in convention implementation work. In the 19th Conference of the Parties to the Montreal Protocol in September 2007, the former State Environmental Protection Administration was awarded "Excellent Implementation Prize" of Montreal Protocol issued by the United Nations and the General Administration of Customs and the Beijing Organising Committee for the Games of the XXIX Olympiad (BOCOG) were awarded the "Public Awareness Prize" respectively. After continuous efforts, China completed elimination of chlorofluorocarbons (CFCs) and Halons, the two most important ozone depleting substances, on July 1st 2007, 2.5 years earlier than provided in the Montreal Protocol. On this basis, China successfully eliminated the production and use of carbon tetrachloride and methyl chloroform on January1st, 2010. In the past 20 years, China totally eliminated 100 000 tonnes of production and 110 000 tonnes of consumption of ozone depleting substances, accounting for a half of the total eliminated quantity in developing countries, and successfully completed the phased implementation task set in the Montreal Protocol. As the biggest producing and consuming country of ozone depleting substances among developing countries, China has not had any non-implementation situations and has become an implementation example for developing countries in Clause 5.

In order to accomplish the medium term objective and the ultimate objective set in the Montreal Protocol, China has taken the following measures: establish effective implementation mechanism, mature management framework, high quality policy law and regulation system, and powerful implementation team to enhance implementation ability. Regarding domestic implementation management framework construction, the State Council approved the National Ozonosphere Protection Leading Group led by the former State Environmental Protection Administration, composed of 18 departments in 1991, and established the Ozone Depleting Substances Import and Export Management Office in 2000 to practically strengthen the co-ordination and guidance for implementation work. The State Council issued the China National Programme to Phase Out Substances that Deplete the Ozone Layer in 1993 and revised this national programme in 1999 to further improve the policy measures for implementation work. Concerning the formulation of policies, laws and regulations, China issued and implemented more than 100 related policies, laws and regulations in the past 20 years, established and improved the ozone depleting substances production, consumption, product quality and import and export management system and developed the system of laws, regulations and policies with the core of forbidding new construction, reconstruction and expansion of production lines and implementing the production, consumption, import and export quota license system. The Regulation on the Administration of Ozone Depleting Substances put into implementation on June 1st 2010 further standardises the management of ozone depleting substances, reinforces punishment for illegal acts and provides powerful legal guarantee for sustainable implementation. In exploration of implementation management model, China proposed the "Four Synchronism" implementation management framework including production reduction, consumption conversion, substitute development and policy, law and regulation construction very early and has obtained good effects through implementation in industrial mode. China is the first practitioner of "industrial mode" and by organising and implementing 25 industrial plans, it included all enterprises in the industry into the overall elimination plan. This avoids the chronic and stubborn disease of "treatment at point, destruction in area" and "treating while polluting", gives full play to the role of competent industrial authorities and trades society, enhance implementation flexibility and independence, reduces elimination cost and ensures due completion of implementation tasks. In December 2011, the Ministry of Environmental Protection held the Kick off Meeting for Accelerating the Implementation of HCFCs Industry Elimination Plan, indicating that China's HCFCs elimination management plan for consumption industry was formally initiate and implemented.

NO.41

Recommendations

Prepare a coherent national plan on climate change which draws climate-related activities currently underway and planned together to improve their collective efficiency and impact.

Implementation

China has worked out the continuous national plan for climate change: in June 2007, the National Development and Reform Commission issued the China National Climate Change Programme which defines the specific goals, basic principles, major fields and policy measures for addressing climate change in 2010. According to this programme, China will take a series of legal, economic, industrial and technical measures to mitigate greenhouse gas emission and increase the ability to address climate change. Since 2008, the State Council Information Office published the white paper of China's Policies and Actions for Addressing Climate Change that introduces the impact of climate change on China, China's policies and actions for mitigating and addressing climate change as well as China's system and mechanism building for this. The white paper is published every year afterwards.

In November 2009, the Chinese Government published the action goal for controlling greenhouse gas emission and decided to reduce CO_2 emission per unit GDP by 40%~45% in 2020. The Outline of the 12th Five-Year Plan for National Economic and Social Development of the People's Republic of China proposes that in 2015, energy consumption and CO_2 emission per unit GDP should be reduced by 16% and 17% respectively in 2015.

On 12 December 2011, the State Council printed and issued the Work Plan for Controlling Greenhouse Gas Emission in the 12th Five-Year Plan Period which proposes that in the 12th Five-Year Plan period, CO_2 emission per unit GDP should be reduced greatly and in 2015, CO_2 emission per unit GDP in the country should be reduced by 17% compared with 2010. Control CO_2 emission from non-energy activities and emission of greenhouse gases such as methane, nitrous oxide, hydrofluoric carbide, perfluocarbon and sulfur hexafluoride. Further improve the policy systems and mechanisms for addressing climate change, basically establish the greenhouse gas emission statistical accounting system and gradually develop the carbon emission trading market. Carry out low carbon test and demonstration and disseminate a group of low carbon technologies and products with good emission reduction effect.

NO.42

Recommendations

Strengthen efforts to protect and improve water quality in coastal and adjacent regional seas by land-based sources pollution control, and upgrade environmental management regulations and government oversight in the aquaculture industry.

Implementation

In order to control pollution from land-based sources and protect and improve marine environmental quality including major sea areas, the system of laws and regulations for protecting ocean and coastal zone environment has been gradually established. The Law of the People's Republic of China on Prevention and Control of Water Pollution revised in 2008 provides a basis for controlling pollution from land-based sources and defines management key points for pollution from major sectors including industry, agriculture and especially aquaculture. In 2008, China formulated industrial guides for refuse management in fishing, commercial ocean transport and passenger transport ships etc. The Regulation on the Prevention and Control of Vessel-induced Pollution to the Marine Environment was passed in the State Council 79th Executive Meeting on September 9th 2009 and implemented on March 1st 2010.

To execute the specific requirements of various laws and regulations, the Chinese Government proposes the marine environment governance policy of "Unified Planning for Sea and Land and Giving Equal Attention to River and Sea" and puts forward the protection strategy of rehabilitating rivers, lakes and seas. The Chinese Government started to implement the action plan for cleaning up the sea in 2001 and initiated the preparation work of China National Programme of Action for the Protection of the Marine Environment from Land-Based Activities in 2006 to combine national objects with global and regional objectives closely. At the same time, the Ministry of Environmental Protection, jointly with respective ministries and commissions continuously carried out joint law enforcement inspection for the protection of marine environment in the country. By the end of 2010, the subjects of joint law enforcement inspection were expanded to 9 departments including Ministry of Environmental Protection, National Development and Reform Commission, Ministry of Supervision, Ministry of Finance, Ministry of Housing and Urban-Rural Development, Ministry of Transport, Ministry of Agriculture, State Ocean Administration, and All-Army Environmental Office. The performance of annual law enforcement inspection work provides a basic guarantee for implementing various laws, regulations and policy measures for protecting marine environment and reinforcing the protection of marine environment.

China's 11th Five-Year Plan for Environmental Protection definitely proposes that the main objective of ocean environmental protection is to prepare and implement emission reduction of land-sourced pollutants as key point, treat and control pollution in major sea areas as breakthrough, strengthen marine ecological protection, enhance the emergency response ability for marine environmental disasters and improve marine ecological system service functions. On the basis of the 11th Five-Year Plan, the National 12th Five-Year Plan for Environmental Protection proposes the requirements for protecting marine biodiversity and requires establishing the marine environmental monitoring data sharing mechanism to provide basis for thorough and deep performance of marine environmental protection. In May 2012, the Ministry of Environmental Protection, National Development and Reform Commission, Ministry of Finance and Ministry of Water Resources jointly released the Notification on Printing and Issuing the "Plan for Prevention and Control of Water Pollution in Major Drainage Areas (2011-2015)" (HF[2012]No.58). Regarding control of land-sourced pollutants, they definitely propose that discharge-prohibited areas and dumping-prohibited areas should be divided in accordance with the marine function zones and the nonconforming pollution discharge outlets and dumping areas should be closed down within the prescribed time limit. At the same time, comprehensive treatment and rectification in coastal areas is strengthened to promote ecological rehabilitation in offshore areas.

Presently, the Chinese Government is organising the preparation of the 12th Five-Year Plan for Pollution Prevention and Control in Offshore Areas, which is the first national sea area pollution prevention and control plan in China.

NO.43

Recommendations

Integrate environmental considerations systematically into China's development and co-operation programme.

Implementation

Environmental problems have become an important aspect considered by China in formulating and implementing development co-operation plans. The Ministry of Commerce has gradually adjusted China's commercial development strategy since the 11th Five-Year Plan period. First, in development concept, it emphasises the requirements for foreign capital introduction quality and structure and that the foreign capital attraction evaluation system that meets the requirements for scientific concept of development should be established in terms of technical content, domestic counterpart funding proportion, resource consumption, environmental protection and employment increase. Second, in capital attraction structure, it encourages the introduction of resource saving foreign-funded enterprises that have strong technical radiation ability and employment increase ability and encourages investment of foreign capitals in agriculture, high-tech industry, infrastructure, new energies, environmental protection and service industries. Third, strengthen supervision and management, regulate capital attraction order, establish the foreign investment monitoring system based on national industrial safety and reinforce foreign-funded enterprises' social responsibilities and professional ethics. In February 2002, the Ministry of Commerce and the Ministry of Environmental Protection jointly issued the Circular on Strengthening Energy-conservation and Environmental Protection

Statistics of Foreign Investment which clearly provides that when transacting the establishment and change of foreign-invested enterprises, all local competent commercial authorities should require enterprises to submit environmental impact assessment documents to environmental protection authorities for examination and approval as well as environmental protection indices. Concerning foreign co-operation, China has implemented the one-vote veto system for enterprises that do not conform to environmental policies and admittance standards, e.g. Xiamen Dimerthylamine (PX) Project, and Shandong Dongying Dupont Titanium White Powder Project.

Regarding international convention implementation, China has actively acceded to more than 30 international environmental conventions and signed bilateral environmental protection co-operation agreements or memorandums with over 50 countries, thus playing an increasingly important role in global environmental protection. To further increase international convention implementation ability, China is preparing the 12th Five-Year Special Plan for Implementing International Environmental Conventions and the 12th Five-Year Special Plan for Reducing and Controlling Persistent Organic Pollutants (POPs).

With regards to bilateral and multilateral environmental development co-operation, the Chinese Government develops and carries out co-operation projects in a planned way in environmental protection ability building, environmental laws and regulations and policies, water environmental management, atmospheric pollution control, ecological protection, chemicals pollution prevention and control, and global environmental problems management and addressing. In the course of co-operation project development, it has preliminarily established the co-operation direction framework with consideration of domestic environmental protection need and preponderant fields of co-operating countries. For example, Italy is in a leading position in environmental technology and monitoring in the world and co-operation projects mainly involve air quality monitoring, energy saving technology and environmental protection technology; the United States and Germany are advantageous in environmental policy and management, environmental legislation and law enforcement and co-operation projects mainly involved in environmental policy consultation for purpose of providing reference in formulating domestic environmental laws and regulations; Australia has unique experience in water quality management, and co-operation with Australia mainly involves water environmental management including trans-administrative region watershed management and ecological compensation etc.; Sweden is full of experiences in environmental management field, and co-operation between China and Sweden mainly involves environmental management; Norway has more studies in global environmental problems, and present co-operation with Norway mainly focuses on mercury pollution control and addressing climate change. In addition, China and the EU are co-operating to implement the Sino-European Biodiversity Protection Project to vigorously advance biodiversity protection work at the local level. Regarding emergency response to sudden environmental events China has successively co-operated and exchanged with the United States, Germany and Russia.

In order to promote exchange in the environment and development field, the Chinese Government established China Council for International Co-operation on Environment and Development as early as in 1992. The Council consists of Chinese and overseas high level personages and experts in environment and development. Now Vice Premier Li Keqiang is the chairman, Ministry Zhou Shengxian of the Ministry of Environmental Protection and President Biggs of the Canadian International Development Agency are co-chaired. The council provides prospective, strategic and early warning policy suggestions to the executive level of the Chinese Government and policy makers at all levels in exchange and propagation of successful experience in international environment and development field and for major issues in the environment and development field and plays a positive role in advancing China's implementation of the sustainable development strategy and building the resource saving and environmentally friendly society.

NO.44

Recommendations

Translate the energy intensity improvement target into more ambitious energy efficiency targets in all sectors; use a mix of instruments to achieve them, including pricing policies, demand management, introduction of cleaner technologies, and energy-efficient buildings, houses and appliances.

Implementation

Since the 11th Five-Year Plan period, the Chinese Government has been enhancing energy saving and emission reduction, and in addition to energy consumption intensity indices, China has increased evaluation for energy efficiency indices. In the 11th Five-Year Plan for Energy Development, China proposes specific control requirements for the main products and energy consuming equipment in the major energy consuming industries. The Outline of the 12th Five-Year Plan for National Economic and Social Development proposes energy resource efficiency indices in addition to energy consumption intensity indices, including proportion of non-fossil energy resources in primary energy consumption, resource productivity, and comprehensive utilisation of industrial solid waste. In departmental and provincial level plans, further evaluation is strengthened for energy efficiency indices.

To ensure the accomplishment of energy saving and emission reduction goals, the Chinese Government takes a series of measures and means. It adopts such comprehensive means to promote the accomplishment of the energy saving and emission reduction goal as accelerating the construction of energy saving and emission reduction project, deeply advancing industrial structural readjustment, eliminating backward production capacities in electric power, iron and steel, building materials, electrolytic aluminum, iron alloy, calcium carbide, coke, coal and plate glass industries and enhancing operation supervision and management of high energy consuming and high pollution industries. At the same time, it implements differential electricity price system, organises special inspection for energy saving and emission reduction work in high energy consuming and high pollution industries and cleans up and corrects preferential policies in electricity price, land price and taxes and expenses for high energy consuming and high pollution industries in all regions. Carry out information disclosure for heavy pollution industries, strengthen the whole process supervision and management for energy consumption quota standards in the new built buildings and implement special monitoring and evaluation for building energy efficiency. Since 2008, energy consumption and energy saving measures etc. should be written in sales or purchase contracts when newly built commercial buildings are sold. China has established and improved the large public building energy saving operation supervision and management system.

Accelerate research, development, demonstration and dissemination of energy saving and emission reduction technologies. Arrange a group of major energy saving and emission reduction technical projects in the key national basic research and development plans, national scientific and technological support plan and national high-tech development plan to accelerate demonstration and dissemination of energy saving and emission reduction technology industrialisation. In major industries such as iron and steel, nonferrous metals, coal, electric power, petroleum and petrochemical, chemical, building materials, textile, paper making and building industries, disseminate the high potential and wide application major energy saving and emission reduction technologies.

The Comprehensive Work Plan on Energy Conservation and Emission Reduction during 11th Five-Year Plan Period released by the State Council proposes comprehensive policy means and tools in target decomposition and evaluation, industrial structure readjustment and optimisation, recycle economy development, improvement of economic incentive policies, acceleration of research, development, demonstration and application of energy saving and emission reduction technologies, dissemination of marketisation mechanism and means including demand management and enhancing energy saving and emission reduction basic ability building. Now various means are being improved and strengthened at all levels.

NO.45

Recommendations

Bolster the adoption of cleaner fuels (including cleaner coal technology, coal washing and flue gas desulphurisation) and cleaner fuels for vehicles, as well as cleaner cars.

Implementation

The National Implementation Plan for Clean Energy Actions issued in 2003 proposed a series of measures to promote the application level of clean energy technologies. The Outline of the 11th Five-Year Plan for National Economic and Social Development of the People's Republic of China proposed restrictive indices that in the 11th Five-Year Plan period, energy consumption per unit GDP should be reduced by 20% and total emission of major pollutants reduced by 10%, which promoted the use of clean energies; at the same time, it definitely proposed that the proportion of non-fossil energies in primary energy consumption should be increased from 8.3% to 11.4%. Regarding flue gas desulfurisation, China's 12th Five-Year Plan for Total Emission Reduction of Pollutants definitely provides that FGD plant should be installed in all coal fired power plants, which puts forward compulsory requirements for desulfurisation efficiency of all coal fired facilities. Regarding specific facility reduction measures, China applied technical means including clean coal technology to increase energy utilisation efficiency and fuel conversion factor.

In the 11th Five-Year Plan period, the Chinese Government took financial and tax policies measures such as implementing Passenger Vehicle Fuel Consumption Limits and encouraging purchase of light cars and thus the energy saving technologies and products such as advanced internal combustion engine, high efficiency transmission, lightweight materials, entire car optimisation design and ordinary hybrid power were disseminated greatly and average automobile fuel consumption dropped significantly. For motor pollution emission in the 12th Five-Year Plan period, the Chinese Government works out the 12th Five-Year Plan for Development of Strategic Emerging Industries and the Energy Saving and New Energy Automobile Industry Development Plan, both of which emphasise and propose to accelerate the fostering and development of the new energy automobile industry and mainly promote the industrialisation of pure electric automobiles and plug-in hybrid vehicles; all out dissemination of energy saving vehicles and enhance the overall technical level of the automobile industry for the purpose of rapidly reducing automobile fuel consumption. Furthermore, in order to strengthen motor vehicle pollution prevention and control, China has formulated the Stages III, IV and V emission standards for heavy duty vehicles and Stages III and IV emission standards for light duty vehicles and is now formulating the Stage V emission standard for light duty vehicles. In order to enhance vehicle fuel cleanliness levels, China has formulated Stages IV and V vehicle gasoline and diesel harmful substance control standards.

NO.46

Recommendations

Develop and implement a national transportation strategy that recognises the environmental externalities of transport and takes an integrated approach to private and public transport; streamline the institutional framework for developing sustainable transport systems; use a mix of regulation and economic instruments (*e.g.* taxes) to give citizens incentives for rational transport decisions.

Implementation

To drive sustainable transport, the Chinese Government has taken many policy measures. First, it established and implemented the public transit priority development strategy; second, it advanced construction of the urban rail transit network including light rail, metro and trolley car; third, it actively developed ground bus rapid transit system to increase traffic network density and station coverage; fourth, it regulated the development of urban taxi industry, reasonably guide travel of private motor vehicles and advocate non-motor vehicle travel.

In the 12th Five-Year Plan period, the Chinese Government strengthens energy saving management and pollution control for traffic transport industry. The 12th Five-Year Plan for Development of Traffic Transport Industry published by the Chinese Government proposes the control targets such as reduction rate of energy consumption and CO_2 emission per unit transport mileage of operating vehicles, reduction rate of energy consumption and CO_2 emission per unit transport mileage of operating ships, reduction rate of energy consumption and CO_2 emission per tonne kilometer of civil aviation transport, reduction rate of land use per unit mileage of national and provincial highway, trafficability increase rate per unit length dock coastline of coastal ports and reduction rate of emission intensity of major pollutants such as Total Suspended Particulates (TSP) and Chemical Oxygen Demand(COD), and proposes specific requirements for energy saving and emission reduction targets of the traffic transport industry. The 12th Five-Year Plan for Energy Saving of Railways published by the Chinese Government definitely proposes that that main energy saving targets for the railway industry in the 12th Five-Year Plan period are: basically complete industrial energy saving and emission reduction laws, regulations, policies and standards; establish and improve the industrial energy saving monitoring system; and reduce comprehensive energy consumption for unit transport amount by 5% from 5.01 t standard coal/1 000 000 conversion tonne kilometers in 2010 to 4.76 t standard coal/1 000 000 conversion tonne kilometers in 2015.

NO.47

Recommendations

Strengthen mass transport in urban areas, and take measures to encourage the urban use of cleaner transport modes (*e.g.* bicycles).

Implementation

The Chinese Government attaches great importance to urban traffic development and energy saving and emission reduction. Since the 11th Five-Year Plan period, in order to implement the "Green Transport" strategy, it implements the strategy of developing public transit and large capacity transit and encouraging the development of clean energy automobiles. It has issued the economic subsidy policy for replacing old cars with new ones and accelerated the update and elimination of high pollution vehicles. At the same time, in order to reduce urban traffic congestion and urban traffic pollution, Beijing, Shanghai and Guangzhou implemented the more stringent motor vehicle emission standard and vehicle fuel standard in advance and issued motor vehicle license plate auction/lot number policy. Beijing and Shanghai enforced differential parking fees and increased the parking fee standard for traffic congestion areas such as city centre.

The Comprehensive Work Plan on Energy Conservation and Emission Reduction during 12th Five-Year Plan Period released by the State Council proposes to advance energy saving and emission reduction for traffic transport, by: accelerating the construction of integrated transport system and optimising transport structure; actively developing urban public transit, scientifically and rationally allocate various urban transport resources and advance urban rail transit construction orderly; increase the proportion of railway electrification; carry out urban pilot work for low carbon transport system construction, deeply put in place special actions of low carbon transport for 1 000 enterprises in vehicles, ships, roads and ports and popularise trailer pick-up transport on roads; comprehensively implementing the electronic toll collection system; realising inland river ship type standardisation; optimising seaway and airway and promote energy saving and emission reduction in aviation and oversea shipping trades; conduct energy saving reconstruction in airports, docks and stations, accelerate the elimination of old automobiles, locomotives and ships, basically eliminate yellow sticker vehicles registered for operation before 2005 and speed up the improvement of vehicle fuel quality; implement Stage IV motor vehicle emission standard and gradually implement Stage V emission standard in the major cities and regions where conditions permit; comprehensively carrying out motor vehicle environmental protection mark management, explore the total retention of regulated and controlled motor vehicles in cities and actively popularise energy saving and new energy automobiles.

NO.48

Recommendations

Pursue efforts to provide the rural population with safe water supply and sanitation to meet domestic objectives and international commitments (*e.g.* Millennium Declaration and WSSD); continue to install meters and collect user charges, taking account of social aspect.

Implementation

In the 11th Five-Year Plan period, rural water supply and sanitary facilities in China were developed markedly. From 2006 to 2008, the central government arranged 23.8 billion CNY and local regions raised 19.5 CNY counterpart funding to solve the drinking water safety problem for 108.66 million people; at the end of 2008, the central government expanded domestic demand and added 100 billion CNY investment, 5 billion of which was used for rural drinking water safety project and solved the drinking water safety problem for 1.571 million rural people; by the end of 2010, China cumulatively solved the drinking water safety problem for 215 million rural people.

In the 12th Five-Year Plan period, National Development and Reform Commission, Ministry of Water Resources, Ministry of Housing and Urban-Rural Development jointly issued the Development Plan for Water Conservancy (2011-2015), which proposes to solve the drinking water safety problem for 298 million rural people and 114 000 rural schools. The Notification of the State Council on Printing and Issuing the 12th Five-Year Plan for National Basic Public Service System proposes to strengthen the construction of drinking water sanitary supervision and monitoring system in the 12th Five-Year Plan period.

The National 12th Five-Year Plan for Environmental Protection emphasises conducting rural drinking water source area investigation and evaluation and advancing rural drinking water area protection zone or protection range delimitation work by: strengthening comprehensive environmental treatment and rectification of drinking water areas; establishing and improving environmental supervision and management system for rural drinking water areas and increase low enforcement inspection; carrying out environmental protection promotion and education and enhance rural residents' water resource protection awareness; and carrying out urban and rural water supply integration in the regions where conditions permit.

Water charge collection is used as an important means to strengthen water resource management and increase water pollution treatment efficiency and is being expanded and applied in the whole China. By the end of 2010, most urban householders in China had installed a water meter. Now as an important construction item for basic service equalisation, water meter installation is gradually being extended to the broad rural areas. Regarding water resource fee and sewage disposal charging policies, due to regional difference of economic development level and development stage in China, the east developed regions are gradually establishing the market-oriented cascade water price system while in the central and west regions as well as underdeveloped regions, the Chinese Government implements the water price subsidy policy.

NO.49

Recommendations

Promote sustainable forest management through issuance of forest management plans, certification of foresting practices, and labeling of forest products in China; expand co-operation with supplying countries in the forestry sector, to ensure that imported wood and wood products are sourced from forests that are managed on a sound, sustainable basis.

Implementation

China is taking positive measures to make sustainable forestry management. In August 2007, the State Forestry Administration, National Development and Reform Commission, Ministry of Finance, Ministry of Commerce, State Administration of Taxation, China Banking Regulatory Commission and China Securities Regulatory Commission jointly issued the Key Points of Forestry Industry Policies. Its policy target is to "comprehensively implement scientific concept of development, implement the ecological construction-based forestry development strategy, exert the fundamental function of market allocated resources and national macroscopic readjustment and control function, gradually establish class-complete, high quality, high efficiency, orderly competing and vigorous modern forestry industry system, sufficiently exert many functions of the forestry, all out promote forest product supply capacity and satisfy the diversified needs of forest products and services by economic and social development to the greatest extent.

In August 2007, the State Forestry Administration and Ministry of Commerce jointly prepared and issued A Guide on Sustainable Overseas Silviculture by the Chinese Enterprises to require Chinese enterprises engaged in silviculture activities abroad to strictly execute the related laws and regulations in the host countries, protect forest lands according to the law, strictly protect high value forests and never change forest land usage illegally.

In May 2008, the Chinese and US Governments signed the Memorandum of Understanding for cracking down illegal logging and related trade and established the Sino-US bilateral forum for cracking down illegal logging and related trade composed of China State Forestry Administration, Ministry of Foreign Affairs, Ministry of Commerce and General Administration of Customs and the Office of the U.S. Trade Representative, Department of State, Department of Justice, Fish and

Wildlife Service, Customs and Border Protection and Forest Service.

On May 5th 2009, the State Forestry Administration and Ministry of Commerce jointly issued A Guide on Sustainable Overseas Forests Management and Utilisation by the Chinese Enterprises to propose regulatory requirements for forest resource management, wood processing and transport, personnel training and technical guidance, establishment of multi-stakeholder publication and consultation system, strengthening of environmental protection and biodiversity protection and promotion of local community development.

NO.50

Recommendations

Review price levels for energy, water and other natural resources to ensure the value reflect their scarcity and internalise externalities; consider mechanisms to compensate or mitigate their impact on poorer sections of the population and regions that would be adversely affected by such price increases.

Implementation

The Chinese Government attaches great importance to the formulation and implementation of natural resource pricing policies to better reflect their scarcity value and reduce their environmental externality. In recent years, the Chinese Government has been addressing itself to advancing resource tax reform. A Resource tax reform pilot was initiated in Xinjiang in June 2010. By the end of 2010, resource tax reform pilots had been extended to 12 west provinces and resource taxes for crude oil and natural gas changed from quantity-based collection to price-based collection. In September 2011, the State Council issued the Decision of the State Council on Amending the Provisional Regulation of the People's Republic of China on Resource Tax, which defines a collection method for tax amount payable of resource tax; in October the State Council issued the Detailed Rules for the Implementation of the Interim Regulations of the People's Republic of China Concerning Resource Tax, which indicated that the resource tax reform with the core content of changing quantity-based collection of resource tax amount to price-based collection began to spread to the whole country from the pilot regions, thus further improving the resource product price formation mechanism.

For the narrow tax item range, unscientific collection methods and low tax burden problems existing in resource taxes, some policy documents such as the Comprehensive Work Plan on Energy Conservation and Emission Reduction during 12th Five-Year Plan Period developed by the Chinese Government also propose to continue to advance resource tax reform and expand the resource tax reform implementation range. In conclusion, China will speed up resource tax reform, increase the tax rate, widening the collection range including the allocation of more resources.

The Chinese Government attaches great importance to water resource pricing policy reform. On July 6th 2009, the National Development and Reform Commission and Ministry of Housing and Urban-Rural Development jointly issued the Circular on Appropriately Fulfilling the Issues Regarding the Administration of Prices of Urban Water Supply to strengthen urban water supply price management; in December 2012, the National Development and Reform Commission issued the Supervision and Examination Measures of Urban Water Supply Pricing Cost (Trial Implementation) to strengthen examination of urban water supply pricing cost and it also issued the Guiding Opinions on Cost Disclosure Pilot Work of Urban Water Supply Price Adjustment to speed up urban water supply price reform pilot work. Water price is adjusted up in most regions in China such as Shanghai, Chongqing, Jiangsu, Shandong and Ningxia to solve the hang upside down of water supply cost and water price and the water price fails to reflect the water resource scarcity and water treatment cost.

The Chinese Government pays much attention to the impact of water price increase on low income people and the population in poor areas. In the Circular on Appropriately Fulfilling the Issues Regarding the Administration of Prices of Urban Water Supply, the Chinese Government requires listening to opinions from all social circles during water price reform to increase water price decision-making transparency and strictly fulfilling water price adjustment procedure and implementing water price hearing system; it requires all regions to give sufficient consideration to the bearing capacity of low income families during water price adjustment, practically to guarantee work for low income families, reduce impact of water price adjustment on low income families and in the light of total price level change, take several modes such as increasing low income standard and increasing subsidy for low income families in accordance with local conditions and ensure that their basic living standard is not lowered.

NO.51

Recommendations

Continue to assign high priority to domestic and regional anti-desertification efforts.

Implementation

China is making efforts to prevent national and regional desertification. In March 2007, the National Desertification Combating Conference was held in Beijing. Wen Jiabao, the former premier of State Council and Vice Premier Hui Liangyu attended the conference and delivered important speeches. At the conference, the State Forestry Administration signed the desertification prevention and control target responsibility document in the 11th Five-Year Plan period with 12 provinces and autonomous regions whose prevention and control task is arduous, to further reinforce the responsibility system of the people's governments at all levels in major provinces and autonomous regions and define desertification prevention and control work tasks and objectives and the measures to be taken and responsibilities to be borne in the 11th Five-Year Plan period. In the conference held by the Committee for the Review of the Implementation of the United Nations Convention to Combat Desertification in 2007, China made a special introduction to desertification monitoring experience and the UN Food and Agriculture Organization chose China as one of the six demonstration countries for global land deterioration evaluation project. In August 2007, China's provincial level desertification and sandy land monitoring information management system was built and put into use. Since May 1st 2008, the first national standard Technical Specification for Combating Desertification in the desertification combating industry in China was put into implementation. In March 2009, the fourth countrywide desertification and sandification monitoring was initiated and the monitoring range covers 30 provinces (autonomous regions and municipalities directly under the Central Government) and about 900 counties and more than 4 700 000 km^2 area. In January 2011, the Chinese Government published this monitoring result. By the end of 2009, national desert land area was 2 623 700 km^2 and sandy land area was 1 731 100 km^2, accounting for 27.33% and 18.03% total national territory respectively. Related departments devised plans and increased implementation. National desert land area was reduced by 2 491 km^2 annually on average and sandy land area reduced by 1 717 km^2 annually on average. Monitoring shows that land desertification and sandification in China have been preliminarily brought under control on the whole.

The State Forestry Administration has printed and issued the National Plan for Construction of Comprehensive Demonstration Areas for Combating Desertification (2011-2020), which proposes the guiding ideology for a demonstration area construction in the 12th Five-Year Plan period and a future period of time. It will adhere to the scientific prevention and control, comprehensive prevention and control, and legal prevention and control policy, and by comprehensively advancing and accelerating demonstration area construction, make efforts to realise a 70% treatment rate of the treatable sandy land in the demonstration areas within 10 years and build a group of desertification combating demonstration and model projects in ecological improvement, technical innovation, policy mechanisms and industrial development.